Philosophic
Hermeneutics

Philosophic Hermeneutics

To order additional copies of this book, contact:
Xlibris
1-888-795-4274
www.Xlibris.com
Orders@Xlibris.com
539923

Introduction

In "Philosophic Hermeneutics" author Wayland Porter coins cutting-edge signage for students of philosophy and philosophical hermeneutics. The book features entertaining bon mots that rely on a sanguinely ironic view of modern life and its technologically abstruse verbiage and foibles. Exegesis and hermeneutics may have influenced such a state and may have had a hand in starting such complexity of thought and communication. Porter de-scales the complexity into verbally attractive sound bites—gambits that may hook in modern youth and yeomanry.

Gnomic and aphorist, "Philosophical Hermeneutics" accesses the various riddles, conundrums and contradictions of modern phenomena. Porter has a charmed and witty manner that makes Martin Heidegger's highly intellectual theory of knowledge easier to digest. Although Porter's sayings may seem wreathed in intellectual mystery and sometimes hews closer to Lewis Carroll's non-sense rhymes, it is the seeming chaos of universal order and the strongest emotions, the philosophical trends of the human herd (a contradiction in terms) which makes them so. Amazing verbal wordplay, they are the keys that may unlock modern philosophy for the young. They will further lead into the rigors of semiotics and the most detailed, philosophically complete exegesis of thought and language known today.

"Philosophic Hermeneutics" is the volume that descends into popular culture and modern society and translates them into the most sanguine, un-complex, yet truly intellectual gambits in the strategic game of human thought and its continuing progress.

For additional copies log on to www.Xlibris.com

About the Author

Wayland Porter was born in Barre, Vermont, and lived in Illinois. He was involved in the business sector for 28 years where he became aware of different economic trends and patterns.

Section One

The Rabbit Hole

http://www.youtube.com/watch?v=9_vYz4nQUcstirytuee

Heidegger speaks, you-tube 1964-69

The notion of language as an instrument of information urges nowadays to the extreme. The relation of human being (s) to language is undergoing a transformation, the consequences of which we are not yet ready to face. The ongoing of this process cannot be stopped by direct intervention besides its going on in the profoundest silence. Evidently, we have to state, that language in everyday life appears as a vehicle for understanding and will be used as such a vehicle. But, there are other relations to language than the common ones. Goethe calls these other relations the deeper ones and says of language "In normal life we make language work in a provisional way, because we are signify just superficial relations. As soon as we speak of deeper relationships, there comes up suddenly another language' that of the poetical.

*

Existential dualism assumes to share the same perception of culture as impressioned by national education and media; "hermeneutics" intuit distinguishes idiosyncratic personifications in comparison in knowledge to any cause in suppositional context disseminating a metaphysical view is quantum physics Quantum mechanics are the means functions which condition human culture as physical mediums in human extensions of forage propentence transfer in any indigenous origins to economies.

The question is what is the question

Context is essential to any comprehension of any object in observation

I Corinthians XIV parallel's Heidegger's observation; see section 2

II Corinthians IV 18

While we look not at the things which are seen, but the things which are not seen for the things that are seen, are temporal; but the things which are not seen are eternal.

Quantum Physics are the cultures conditions to subconscious archive in domesticating conformity

There is what you know in experience and there is what you do not comprehend in any physics to the causes of conscious subjectivity "Eternal assumes intelligent design however eternal is intuit or instinctive natures to corporal functions in survival to adaptation of environment. The basic physics to consciousness in culture are provisional paternal senses of interdependence. Transgression is a natural instinct of survival however transferred to our interdependent material culture entailing virtuous disciplines. As children we do not know how to communicate our challenge in adaptation to our culture. Common sense is any paternal appreciation.

There's no such thing as a collective conscious mind to knowledge, knowledge is selective interpretation as individuals however all people are affected by a paternal interdependence conditioned by law; philosophy is the science of mind to culture. What is paternal literacy in the game of ego and identity in assimilation to the existential condition to material sustainability. Intelligence is paternal is the law of the people is any origins to the transfer to a centralized authoritarian ambiguity. Intimacy is intellectual sharing any sense to value, carnal is an emotional fixation conditioned by insecurities. Law is an invention to control in the auspicious cogence to obedience as an authoritarian whip. Law infers intent however there is no evidence to intelligent design there are no laws to nature only principles to physical existences is the cause to science. Con-science' what are any origins to words? "Cultural relativity" Nationalism is a symbolic affiliation knowing no common cause in the conformity to patriarch memorialization. Nationalism conditions an idiosyncratic persona enchanting any sense of pride is a penitent sense of being. Socialist values concern any balance to capitalisms distortions to common wealth in the sense of equal opportunity where education provides no guarantee to placement in forage sustainability.

Consciousness; there is no cause there is nothing to know you simply exist it is not what you know it is how you know anything is any practice in comprehension. Knowledge knows cultural conditionings are empathetic associations to common senses. Human interdependence is a provisional directory in how technology conditions existential experience;

Memorization scripted in historic knowledge having no construal in relativity conditions an illusory paradigm

Heidegger continued

The decisive experience of my thinking, and that means at the same time for western philosophy the meditation on the history of western philosophy has shown me that in the past one question did never appear: the question of being. And this question is relevant because we determine, in western thinking that man is in a relation to the being and that he exists by corresponding to the being.

The task which is given to the thinking nowadays as I understand it is new in the sense that it requires a new method of thinking, and this method can only be used in the direct dialogue between man and man, and can only be attained through a long exercise and through an exercise, as one might say seeing in thinking. That means; this way of thinking can be performed, for the present, only by a few but can be then, mediated through the different educational spheres, communicated to others. I'll give you an example. Today everybody is able to operate a radio or television set without knowing the laws of physics in their substantial contents are understood today only by five or six physicists. The same is valid for the task of thinking.

All education originates from philosophic interpretation challenging institutional authorities

True physics remain constant the discovery of such things was a process of accumulative research performs only by a few physicists, "Incremental fundamentals to comprehension" understanding physics; the origins to telephone communication was the invention of the telegraph, physics have ancestry, sound is a vibration; the task of thinking of a few is the task of interpretation providing any physics to our culture; Heidegger's concern was education' "how we know what to know" as the means functions expand our conception of the world becomes more significant, knowing how any technology functions is intimidating regarding what one would have to know to be a physicist is any specialization, the same is true for metaphysical science you don't require knowing how technology functions they just do unless you want to be a physicist; in philosophy it is how anything applies.

The question of the demand for a world change leads me back to a famous sentence by Karl Marx in the Theses on Feuerbach, and I will, to be correct read it from the original; "The philosophers have only interpreted [underscore] the world, in various ways; the point is to change it." If we cite this sentence und if we follow this sentence we disregard that a change of the world presupposes a change of the conception of the world. And that a conception of the world will only be obtained by a sufficient interpretation. That means, Marx relies on, on a certain interpretation of the world to demand his change of the world. And therefore this sentence proves to be a non-founded sentence. It provokes the impression that is decisively spoken against philosophy. While in the second part of the sentence unexpressed the demand of a philosophy is presupposed.

I would say that men—for example in communism—have a religion, because they believe in science. They believe unconditionally in modern science. And this unconditional believe in science, that means the confidence in the certainty of the results of science is a belief and it in a certain way— something that exceeds the existence of a single person, and it therefore a religion. And I would say; no man is without a religion, and every man is in a certain manner transcending himself, and that means; ver-rueckt"

*

We use many instruments where only a few physicists know the functions to these technologies where clients simply push buttons where things magically appear, what are radio waves it's hard to imagine how there is some invisible connection floating in the air seems very fascinating wanting to know any comprehension to the physics of vibration is any specialization.

The cause to philosophy is that we have no connection and what is, is that we all know is that we all live in isolation assuming anything is in control. Technological innovation conditions our minds is what Heidegger was expressing not knowing how to describe cultural anthropology and what is it but what technological evolution does expanding the existential to the means functions. Technology conditions economy which he did not know how to interpret expanding education to the various positions in retrospect changes our environment in expectations to fill positions by the advancement to education; the truth is that we have no education to our evolution to what drives direction though somehow things catch up. The truth is that most governance is all reactionary which conditions most things in hindsight. This regards any concept to how education prepares anyone to interpret what to know in advance to fill any position knowing any hindsight to discovery to any physics to technology because there is often no common interpretation disseminating discovery. What Heidegger was saying is that there is no "authority" knows how to describe technological evolution advancing education. His reference to Karl Marx refers to what society does scientifically to provide any interpretation to what enterprises our culture… The premise to the question observes there is no economic constitution to what conditions any economic social solution providing for a better constitution to what provisions any economic placement. Cultural Anthropology is our evolution which is part of our education. Technology can be considered religion people worship innovation.

Questioning the argument; Presupposes a change of the conception of the world will only be obtained by a sufficient interpretation. And therefore this sentence proves to be a non-founded sentence, you cannot change the world one can only provide any interpretation. The interpretation of the world is the cause to philosophy defining Cultural Anthropology.

*

Questions are not happenstance thoughts nor are questions common problems of today which one picks up from hearsay and book learning and decks out with a gesture of profundity questions grow out of confrontation with the subject matter and the subject matter is there only where eyes are, it is in this manner that questions will be posed and all the more considering that questions that have today fallen out of fashion in the great industry of problems. One stands up for nothing more than the normal running of the industry. Philosophy interprets its corruption as the resurrection of metaphysics.

Philosophy, then, is not a doctrine, not some simplistic scheme for orienting oneself in the world, certainly not an instrument or achievement of human Dasein. Rather, it is this Dasein itself insofar as it comes to be, in freedom, from out of its own ground. Whoever, by stint of research, arrives at this self-understanding of philosophy is granted the basic experience of all philosophizing, namely that the more fully and originally research comes into its own, the more surely is it "nothing but" the transformation of the same few simple questions. But those who wish to transform must bear within themselves the power of a fidelity that knows how to preserve. And one cannot feel this power growing within unless one is up in wonder. And no one can be caught up in wonder without traveling to the outermost limits of the possible. But no one will ever become the friend of the possible without remaining open to dialogue with the powers that operate in the whole of human existence. But that is the comportment of the philosopher: to listen attentively to what is already sung forth, which can still be perceived in each essential happening of world. And in such comportment the philosopher enters the core of what is truly at stake in the task he has been given to do. Plato knew of that and spoke of it in his Seventh Letter.

~Martin Heidegger ~End You-tube

Philosophers share notes to what consciousness questions in association prompts context

Postulating a true world view communication often assumes common knowledge following a collective story

What is the conflict in question assuming any authority knows any cause; philosophy provides any contextual interpretation of our culture to common experience in adaptation to material domestication to be used as a guide in confirmation. Philosophic Hermeneutics is the method used in philosophy for providing a common language to cultural anthropology.

Heidegger was so poetic; the thing is that we are all exposed to the same cultures inhibiting subjectivities checking our intelligence is our challenge to make sense of our individual and collective experience; now what conditions us is any assumption to authority knowing any directive in heritage; any future is any abyss when you know how to read between the lines you become a philosopher and then there is what we do keeping things going in any assumption there is any goal wanting to survive in the sea of confusion wanting to be a part of anything within our own gauges of achievement. The thing is is there is no purpose we are simply born into the bondage of our material culture which converts resources into modern conveniences is quantum theory resolved. Life is a struggle for most people who do not have the reserves in financial backing providing any support so what is reality is what capitalism does corrupting any mind and yet there is such a thing as economic interdependence. So what is reality if you do not know how to apply social consciousness to virtue in being one mind assuming any interpretation to democracy? The world leaders are people too asking the same question; how can we control human nature. The world view is what you know around you. You sense pretention when you are surrounded by a group of people contemplating what anyone knows as a world perspective there is the social and economic being one provides the other, what you do for a living is a posterious identity in comparison, the thing is you are vying for your own survival in a precarious world of expectation it is hard to be real without a portion of anxiety.

http://www.youtube.com/watch?v=hkCR-w3AYOE&list=PL2DD5D69B7D254864

Life is worth losing by George Carlin

Human physical attributes in agility pursuits crafty transformations is the primacy to the existential being conditions cultural intuit in holographic reflection

Quantum mechanics; what are the physics to our culture driving the collective bus providing any interpretation to course assuming any authorities; you are all alone in the world until you make any connection knowing what you know to anything in common postulating a collective mind to knowledge. Cultural Anthropology is a psychological profile to an existential embodiment as humans evolving with technology conditions conscious subjectivity precepts comprehension' context is essential for any physical association being construed is a cultural constitution to our existence.

What does anyone know? Reading others thoughts challenges comprehension to our own experience is subjective being that our culture is all characteristic in any appreciation to commonality seeking any connection? Social subjugation is the realization that we do not know how to communicate our true senses to an assumed collective mind, there is no collective mind because there is no interpretation to our culture, you cannot be dedicated to anyone but yourself no one knows who you know and yet we are all the same cultural influence in association to any contribution dedicated to the assumption that we are one union knowing any disposition, there is what you know and there is what you realize when we walk into the world postulating any connection can be reserved; realization is knowing a true connection being that there is no true goal to apply to because there is no purpose or cause only casualness. Now what is a true world anyone knows how to apply to? You want to be a part of anything making sense is the drive in ambition wanting to be successful. There is no collective mind providing any interpretation; sanctuary is a provisional directory to subsistence pulling your own weight knowing you have no control over the whole. Expectation contributes to distractive thought assuming what any one knows which is nothing because we have no clue what our culture is as there is no instruction within how to apply our minds to any definable cause. And then you have to work for a living being a force having no art in the cogence to expectation conditioned to law. So what is reality to anyone where there is no art to discovery? Reality is the expectation of our cultures interdependence unknowing what culture does; there is no authority, authority infers a purpose to any goal there is no cause being that we have no idea what drives the culture. Now what do you know; nothing because there is no interpretive goal other than material procurement within any individual sustainability.

Cultural dialog

Heidegger's documentary was produced in 1964-69 as he was living through a time where people were waking up asking the question what are we doing like any times. A common language is the physics to our culture in the poetical. The sixties were in a depression (over population) trying to fit in to any position in subsistence due to the baby boom, (1945-1960) where there was for many a sense in resolve, there are patterns to economy stimulated by technological innovations providing convenience which expand position to our forage propentence transfers to economy, human physical mediums are the equal origins to our cultural domestication and interdependence; clothing is a medium of the skin, clothing conditions our genetics (We once were hairy beasts protecting us from our climate depending upon temperate zone) being that we evolved from the earths origins (the big bang) unlike other species we have physical attributes giving us abilities to use tools to innovate our cultural conveniences expanding the use of our minds developing our conscious experience often congested in the memorialization to an imperfect world wanting to be successful.

What he was saying is that we have no clue what we are or what conditions being, any physics to the culture are the only veritable education to what we are subjected being intuitive knowledge. No one can change the physics to our culture one can only interpret anything to provide a common language to better understand our interdependent nature, this is what Karl Marx did not know how to do being that he did not understand technologies origins (Marshal McLuhan did) the truth is that no one can dispute true physics. Quantum Mechanics is what drives the bus and has been misinterpreted by philosophy. The question is where did anything begin what does anything do and where does anything go distending the future.

The Hermeneutical Problem in the Sciences by George Gadamer

My own undertaking was itself conditioned by an 'effective history'. Obviously it was rooted in a very definite German philosophical and cultural heritage. The so-called Geisteswissenschaften ('human sciences') in Germany had really never before so completely united in one package their scientific and their world intuitive functions. Or, to put it more bluntly, they had never so fully and consistently concealed their ideologically conditioned interests behind the epistemological and methodological pretensions of their scientific procedures. The indissoluble unity of all human self-knowledge_was expressed much more clearly elsewhere than Germany: in France through the broader concept of 'the humanities'. What I wanted to bring about by insisting on the 'historically affected consciousness' was a correction of self-concept of historical human sciences, and here I include scholarship about art: that they are not 'sciences' in the manner of the natural sciences. But bringing about the recognition of a 'historically affected consciousness' in human sciences was not my goal, for the full dimensions of what I have called 'the hermeneutical problem" are much broader. In the natural sciences, too, there is something like a hermeneutical problematic. Their path, too, is not simply that of methodical, step by step progress. This has been persuasively shown by Thomas Kuhn and was already implied by Heidegger's 'The Age of the world Picture' as well as in his interpretation of the Aristotelian view of nature. Both make clear that the reigning paradigm is decisive for the questions research raises and for the data it examines, and these are apparently not just the result of methodical research. Galileo had already said Mente concipio I conceived through the mind. Galileo worked out his experiments in his mind before performing them physically.

Consciousness is complex and not so complex; the nature of consciousness is conditioned by culture... Most experience has no reflective mediation in understanding how the mind conditions and categorizes identities or characterizations in comparison affects the way we perceive the world as individuals... the observer being observed conditions idiosyncratic polarities in the ego and chameleon paradigms to association...

Having expectations using psychedelics for enlightenment conditions thought thinking there is something to know, there is nothing to know enlightenment is the practice in comprehension in contextual relativity to random or prompted experiences is philosophic hermeneutics.

The mind is in observance of the self is existential duality reasoning experiences in mediation to cultural characterization to normal in the game of ego and identity in comparison subject to classifications conditioned by economic successes in the game of capitalism's pretensions.

Children have vivid imaginations... there is no true world view... kids are just expected to know... what is it? There is no understanding in children of "structure" kids like to play games with their parents or others making things up for attention, our culture is confusing isn't it... Parents assume there is a world view we all know... The view of children has limited experience in domestication or self-reliance which is gradual in adaptation to responsibilities... in childhood intuit is play and time is irrelevant. And then you have parents living in the real world of pressures and the expectations of their children to do what they are told without knowing why anything is relevant... Why is the first intuitive question in children...

The brain does not have anything to do with "intuit" in cognitive function; there is no preceptual operativeness to conscious experience. The brain is conditioned by visual perception in coordination to motor functions to environmental habitat. Manageability is any structure in practice conditioned to financial occupation in a provisional sense in material being

Hermeneutics is a process of understanding experience in conversion to language; knowledge is experience of our culture in adaptation conditioned to the perpetuity of changing technologies expanding and obsolescing economic positions, where time is conditioned any occupations should be an art'

Hermeneutics is a cultural soundex

Physics are simple when you know what you comprehend" conditions poetical determinations

Hermeneutics observes a portrait in time defining the brush strokes to our culture

The key to any comprehension is contextual relativity

Not having any expendable income has a sense of strife and feels degrading

There is no collective dialog there is no venue only religious gatherings and transient causes to liberty restorations

Existential duality assumes we share the same perception to time' perception is conditioned by age in experience

A sense in failure as a parent involving children conditions codependence in guilt often lends to manipulation in control mechanisms to distorted responsibility

Much thought in rehearsal practices desired outcomes

Reality tunnels have kaleidoscopic versions in veritable dispositions

If you think the world view is changing it's your own perception...

Human precognition is primordial instinct in survival subjected to culture

The force in material conformity conditions thought weaved in confusion

A common world view is a pendent gauge in ideological conflict

Memory lanes in social settings have multiple versions in character associations

Losing faculties with drugs in an altered state of mind circumstances criminal intent

Law conditions a liability in restitution in a paternal sense to order postulating a collective social virtue

Forage occupation in transfer to economy conditions an existential presence in sociopathic incrimination

Desperation prisons toil

Respect has a common equality

Speculation has many desired outcomes

A sense in being has a tailored wardrobe

Any act seemingly wrong wants another verdict

Drunkenness wallows in the shadows of unfulfilled desires

Respect has a common equality in virtue

Law telling you what not to do provokes emancipatory retort

Success wants to be known in emancipatory approval

Political hindsight in global irreconcilable dissonance is a politician's virtue

Malevolence to paternal polarity in family setting conditions values in appraisal

Human history has consistency in ideological conflict conditioned to a moral deliberation to resolves

Ego in identity has a posterious dominion in perceptions to reality

Experience in cultural perversions have layered perspectives to influence

Violence is a disgruntled associate

Being who you want to know has interesting spectrums in chameleon characterization

Social exploitation has estranging associations in behavior acceptances

Authority postulates a common knowledge to law to proprieties in corruption

Much speculation analyzes motives in retorts

Greed has no common investment in societal virtue

All things in moderation wade in the changing tides of season's greetings

Invention provides any physical necessity for any proficiency

Theology is theoretical logic to understanding the nature to our culture

Political media symposiums delegate no virtuous resolves to blending cultures

Much speculation causes rhetorical thought

Thinking: Intuit in speculation has an ardent correspondent in reflective mediation in the quandary of being

Law wants a harmonious obedience thou shalt do as ye are told

Illiteracy is the main cause to social distortion... existential polarity postulates a collective mind to knowledge

Graphic violence in movies has a sadistic censor

The first blow has an incriminating seduction in test of courage

Media impersonates a collective mind involvement

Democratic cause for liberty is conditioned by social economic discordance

Desires dream in successes tailors sanctuary in prominence

Liberty is a state of mind not a democratic condition

Provocation has any seductive lure in vindication

Persona has any desire in expectation to perfection

Authoritarian discipline has no intuitive composure

Precedence in illicit behavior has a supercilious dominion

The question is not whether a god exists the question is defining collective intent to authority

Criminal intent has any malice

Dejection has a certain seduction in conquest

Education conditions a custodial polarity government has you covered senses isolation in abandonment post-graduation

Belief in a God assumes an intelligent design in authority to existence

There is no pre conceptual operativeness to conscious experience in knowledge to culture

Time is age in perception to culture

Culture in existential polarity conditions perception in ideological mediation

Social comparative conditions much speculative thought in character association

Character wants a personal cue in association

Authority assumes a collective mind in culture

Practice in comprehension to physical relativities and functions conditions Intelligent Query is the contextual application to education

Authoritarian obedience subservient's competence in repression to individual liberty to mind

A world view is any depth in perceptions to interests

Social characterization conditions an aura in school

Authoritarian virtue has no jurisdiction

Desire wants an intuitive companion

Conquest in betroval wants diplomatic immunity

Reflections in character condition posterious ambitions

Stares have a strange lure in reproach

Discrimination has many characteristics in appearance

Most thought has no resolve

Malevolent provocation trancing corporal seizures wobbles knees

Soap opera drama infirmity consumes maladies

Reality is a cultural malady in ideological conflict conditions polarity

Uncertainty assumes a destination

Inspiration wants gratification

Conscience has a self so virtue

Codependence is an emotional addiction for comfort

Imagination in desires fruition mystical dreams

Addictions have any convenient associations posturing societal exclusion

Being told what to do challenges competence

Law is conditioned to respect to people and property, in polarity intuit knows any seduction

Political left and right is a shadow boxing paradigm to social inequality in economic apportion subject to free trade capitalism and corporate franchise domination

Character wants a compliment

Popularity contest in school has a conceited association in classifications

Manipulation has any contradiction in terms

You want to prove to peers your value as a Good Samaritan gaining restitution

Gods have many ambitious forums in miracle menageries

Humility in sincerity has bigheaded wardens

Youth in the freedom of time has few accouterments

First acquaintance wants an amical evaluation in real time

Formidable years twelve to seventeen adolescent traumas condition perception in value

Financial prosperity conditions immunities to social corruptions

Age in time conditions a memorialized view in authoritarian polarity

Regret wants another rehearsal

Vulnerability has any ideological armor

Ego and identity in comparison condition perception to a class menagerie

Martyring maidens cater male malevolence

Youth has a certain trust an authority knows the culture

Acting wants a caricature expression

Social malevolence has a negative aura

Commitment wants a common denominator

There is no collective mode only segregation to environmental influence in mediations to immunities

Reflective mediation in social perception is a foggy philosophic consultation

Getting high on weed may sense an existential polarity to time/expectation having a paranoiac ambiguity to authority

Human behavior is conditioned by a balance to appreciation in failure and success

Perception in youth has an estranging paternal paradigm in discrimination

Magnetic influences in misdemeanor are embarrassing associations

Control to chaos discretion advised

Law is transgression unless you are innocent

Perception in time and age is a mystical frolic

Intuit in survival is conditioned to an economic material interdependence

Seat belt laws usurp probable cause

Comparison has any social deliberation in paternalistic covenants

Restitution wants anyone to know who did what to whom for emancipation

Perception has a gauge in association to social distortion

Stupidity has a contemptuous nursery

Societal animosity has a martyring disposition

Manipulation wants to convince you of anything you don't want to know

No one knows what anyone knows in perception to associations

Conformity has an estranged association with authority

Aggressive people challenge intuit

People pleasing has an estranged attraction

The mind has a material inventory

Slobs don't make good roommates in custodial occupations

There is no collective conscious reality

Social harmony is a childhood desire

Empirical age in social distortions channel blinders

Character in reflective association has a third party nemesis ward

You do not know anything anyone tells you, you do not know in person

Television sitcoms stage periodic social ambiance

Happenstance acquaintance postulates common perceptions to current tides

Mystic modes in pleasant company want you to be there and what is it

Articulate cognitive computation is a practiced art in any desired interest

Nature's consensual age is self-reliance

Law challenges competence to a paternal intuit conditioned by a domesticating material culture in proprietary possession to contribution associates to an Islamic debate in human rights

The whole concept of knowing what one believes in God as an authority is superstition; the only definitive interpretation to God is "knowledge in culture" which is confusing in the bible within the avowals to its authors "ideology" of creationism which has no evidence.

There is no justice to law in capital punishment postulating restitution is a societal corruption in vengeance as a deterrent is a reprobate parody does not reverse any causes to capitalisms perversions... Authoritarian rule has no intuitive virtue in the philosophy to obedience in subservient polarity to judicious hegemony. Capital punishment is immoral in relevance to a paternal condition to a societal incrimination in the transfer of an ambiguous authority. The question of authority weights the conceptualizations to moralities which are conditioned by a paternal sense to order of conscience in cultural culpabilities.

All children are challenged in language development in experience to desire and emotion

Listening to children trying to express themselves challenges appreciation

Childhood environment sets any precedence in social polarity

Codependence holds opinions of one or another in a paternal malady provoking defenses in assumption there is a shared perception to who anyone knows weights anchors

Plato Meno. *Can you tell me, Socrates, whether virtue is acquired by teaching or by practice; or if neither by teaching or practice, then whether it comes to man by nature, or in what other way?*

Virtue follows intuition in practice... God is one's own knowledge,

http://www.youtube.com/watch?v=ZacggH9wB7Y

Quantum Mechanics (an embarrassment) sixty symbols

What is the interpretation to Quantum Mechanics? Sean Carrol is still a theorist "we as a scientific community do not agree", what is the objective without objective there is no contextual point the point is how is anything predictable.

Quantum Mechanics what is the physical primacy to our culture with what conditions conscious objectivity. Context is the most essential factor to comprehension within any supposition.

Wave function assumes a time space continuum; there are no multiple universes, space is not physical it is space, there are no black holes, now blow up a balloon inside another balloon. Entanglement with environment has nothing to do with electrons assuming a motion picture to time, we all live in our own reality tunnels individually assuming anyone knows our perceptions recording, there is no collective eye only what media presents as relevant. There is no physical time what changes is the physical structure due to technologies evolving changing fashions. The world we know evolves in human image and other species attributes "birds and fish" as physical medium transfers; domesticated accessories to an economic interdependence all originated from the same source as any invention to convenience.

There is no point to all these Neo's out there trying to figure out some code, there is no code there is no collective mind to knowledge only similarities to adaptation in phase to experience in limited reflective mediation. There is no changing collective mind "all there" as the movie Pleasantville fantasies as being one mind. There is only what anyone can proscribe as any physics to our culture, there is no collective mind knowing cause. Now what is a true world view when there is no resolves to anything? There is no true historic recording there is only what the physics to our culture do is inconceivably overwhelming to contemplate. Definitive history is cultural anthropology; there is no true historic accounting to any past to present to what determines casualness. Historic inference is bias and selective so what makes anything of authority. There is no interpretation to our culture so there is no design to follow, design infers intent; what is it assuming anyone knows is any derivations to religious assertions.

If you consider yourself Christian have an open mind this is what the author to the bible did not know how to do, what do Christians know in common with others? It is impossible that there were forty authors it has also been assumed that the bible was written 70 years after the death of Jesus would be impossible being that no one was there writing anything down in real time retrospect assuming there was some massive change to a collective mind which does not exist then or today, no one speaks as they write and no two people wrote in the same style, there was no collective media before the printing press in 1430; the writings in the bible are poetical' summons the realization that it had only one author. The essential theme to the bible is not much different than any artist assuming such a thing as one mind could exist, there neither is nor was or will there ever be one mind, as the Matrix Trilogy also impressions. There is no physical congruence or a common dialog or linguistic constitution. What does anyone know assumes any authorities are a mystical assumption in any concept to being? Our identity becomes what we do in posterity to contribution having no collective interpretation. Everything is control and what is it within any assumption to any outcome that does not exist knowing any directive.

What is reality assuming any connection being all on your own drifting in space asking questions to what I am to anyone? Space and time coordination concerns what you do making sense of the world around us because there is no collective purpose knowing a goal. Identity is conditioned to technology as our language more than we imagine wanting any connection. The cultural epitome is what we all know and do not take the time in reflection knowing the causes to common influence.

Notes:

Section Two

There is something missing in elementary education where memorization rather than contextual comprehension assumes good recall is a practice to "in-tu-it", intelligence is knowing what you do knowing how to know the functions to anything physical, memorizing nouns uses no intuitive deduction sets precedence to what is assumed as an increase to skills in comprehension. Memory recall often knows a story however, how accurate is any people's version when there is no collective eye keeping track of the world's events, only media's investigations construing accounts. Memorizing people places dates and things is non-intuit and only idiosyncratic memory impressioning any memorialization assuming association. Watch the television game show Jeopardy or play the game trivial pursuits. It is not what you know it is how you know anything. Context is fundamental in comprehending anything factual. In-tu-it construes contextual perception. Every mind uses the same fundamental tools, language is physics, and there would be no physics to conscious observance if there were no invention. Invention evolves AI to the means functions conditions cultural interdependence in forage propentence transfer.

"Postulating a world view" much education is non provisional memorialization to nationalist bias categorization initiating idiosyncratic conformities. Adopting any association to culture applies any intellectual objectivity assuming any purpose is applied. Mechanical skills are pertinent knowing how to work through any project systematically is very competitive (let me show you how to do that properly my way) applies as well to politics assuming we all share common determinations to purpose. Memorialization conditions subconscious within any sense of failure, there is no failure, failure is any practice to any perfections to physical and social arts. Human culture transfers forage propentence in the transcendence to technologies conditions communal interdependence to economy is an agricultural toll. Karl Marx did not invent communism he like any philosopher was trying to provide an interpretation to a culture in transcendent evolution. A cultural plane to consciousness conditions a paternal plane is any basis to an interdependent psychology in the realms to judiciary determinations in a paternal sense to order.

Surveillance society is the world which has evolved in human physical mediums; the camera is an extension of the eye, recording devises are an extension of the ear and nose, big brother is a nosy authoritarian. You have every right to speak your mind there is no collective purpose; government is formed by and for people or is it? What is representation? What are you conditioned to that makes sense you feel like a criminal always looking over your shoulder wondering whose watching. A world view is a very precarious assimilation. Cameras do also provide securities to capture those who are engaged in criminal acts provides some assurance that we are being protected is a good deterrent. Authoritarian memorialization has a transient culmination connected to the advancements in technologies. Desperate people do desperate things cultural antipathy stimulates paranoid societies; cultural embodiment induces memorialization wanting ethical members. Now watch Beverly Hills Cop; today movies impression very aggressive portrayals in law enforcement procedures and protocol imploring accepted behaviors is a conforming impression stimulating anarchistic minds not wanting to be controlled wanting to work things out on their own. Assuming anything subjects the mind to any assumption there is a collective mind knowing any directive. There is no definitive world view perspective only where eyes grasp anything in survey Technology is at its culmination to expansion being that all origins began as human mediums, now we build robots and prosthetic accessories to other mediums. .

Linguistic constitutions; an extension of the human mind computers will never be intuitive; they can only be programmed by people transferring scripted associations: the cartoon Transformers is a metaphor. Robo Cops make any sense as prosthetic beings, human narcissism ambitions its culture.

Money is the physical medium extension from forage propentence, as a citizen what a government owes is not the debt of its citizens, who had no control over what government does spending money on space exploration energy projects scientific research or global conflicts; who are the beneficiaries, Corporations. Each country should take care of its own affairs including America. Hollywood movies impression there are many social perversions. Parody does not admonish the causes, what are they?

The reason many people end up in conflicts with the law is not thinking things through. A scrap metal scavenger removes from a creek in the mountains a rusty culvert and is charged with theft III, carrying a $700.00 fine and ten days community service; what is community service if not assuming an obligation to society assuming a contribution to what does that make any sense that he was also being an environmentalist. Law makers have reprobate mind there is no unified code to law. What many people go through to survive is pitiable. If you are homeless and urinate behind a tree are they going to arrest and fine you? When you have lost everything for whatever reasons it is almost next to impossible to find employment. The movies Fisher King and My Man Godfrey are great examples within how capitalism is so perverting like we don't know it. There should be a national recovery program for the homeless sponsored by those in Hollywood who do nothing more than make a living off the worlds social perversions stimulating more distortions. You love actors who simply play themselves over and over while true artists The Munster's Adams Family Beverly Hillbillies are often type cast, today running out of something new movies are often remakes. Please tell us no one is going to attempt a remake on The Pink Panther or Monty Pythons Holy Grail, let us all give them another academy award assuming contribution… The movie "You Can't Take it With You" 1938 is one of the greatest arts ever created.

Largely misdiagnosed by medicine bipolar is relative to cultural aptitude in resentment in a force to material compliance

Why the debate on Religion verses Science persists does consider the social economic malady, the bible is a metaphysical allegory of our culture

You know there are things you can't describe assumed are common knowns we all know the same primacy to our culture's domestication

No one knows what anyone knows you can only share in common a scientific description to our culture

A wise person does not know anything everything is new as Heidegger mentioned that the world is superficial being that communication tries the senses in any expectation is a susceptible antipathy waving appreciation.

Any distortion consigning authority is any assumption there is any cause though law imposes a moral conviction conditioned by a paternal sense to order in its philosophy

The book "The Law" Frederic Bastiat 1848

The complete perversion of the law; pg. 8

But, unfortunately, law by no means confines itself to its proper functions. And when it has exceeded its proper functions, it has not done so merely in some inconsequential and debatable matters. The law has gone further than this; it has acted in direct opposition to its own purpose. The law has been used to destroy its own directive: It has been applied to annihilating the justice that it was supposed to maintain; to limiting and destroying rights which its real purpose was to respect. The law has placed the collective force at the disposal of the unscrupulous who wish, without risk, to exploit the person, liberty, and property of others. It has converted plunder into a right, in order to protect plunder. And it has converted lawful defense into a crime, in order to punish lawful defense. How has this perversion of the law been accomplished? And what have been the results? The law has been perverted by the influence of two entirely different causes: stupid greed and false philanthropy.

The book "The Law" was written during a time in despotism conditioned by fear of economic collapse during recession stimulating greed and uprising in a fear of collapse

Law has no right over personal property or private domain; which includes your own person*"there is no cause there is no authority only the law of the people"* no one has the right to search your home or vehicle without a search warrant, often compelled by authoritarian coercion, *"do you have anything you are hiding you want me to know about"* authoritarian transgression is a physical stressor and criminal in defining criminal as the intent to harm… causes emotional duress much as does being called before a judge. Communism is a police state provoked by transient causes or invasion is illusory regarding cumulative terrorist currents justifying usurpations to civil liberties. War on terrorism is counterproductive where diplomacy reasons ideological causes to common intuit. Cultural divergence challenges indigenous heredity to tradition and "law" concerning human rights.

The law postulates a moral authority transposing its objective in reprisal has no emancipatory restitution repealing any cause in emblematic virtue

Protocol to excessive force in law is a corrupt badge and a perversion to which law is meant to counteract as authoritarian retribution has a transgressive culture.

Describing any conscious state of mind cause or condition in a group or you-tube videos suggests there is anything to know is known there is no cause there is nothing to know it is not what you know it is how you know anything.

Authoritarian despotism in an accumulation to law has any sense in loss to liberty has a culmination sparking civil unrest/rebellions IE "Ferguson MO and others" much like the sixties rebellions. Law is also stimulated by health safety concerns caused by increasing protocols to technology such as seat belt or cell phone use, restrictions or smoking ordinances as authoritarianism is an existential paternal paradigm as a cultural custodian watching over your shoulder. Race ethnicity and economic prejudices play roles in stereotype profiling is repression in the opposite of laws objective for preserving individual liberty and postulating a paternal fraternity of one mind to principalities.

There are cycles to despotism caused by economic patterns to inflation followed by deflation stimulated by advancement in technologies expanding economies positions, conditioned by increases to population capitalism, corporate franchise dilution and consolidation

*

Democracy assumes certain equality in distribution to opportunity and equal rights; despotism on the other hand is a distance in social disparity and equality to sustainability contributing to social trepidations. So how does democracy provision free trade capitalism and what is any actual role to government as an individual and collective transfer in authority where there is no economic constitution…

Religious debates are redundant; IE' law is based upon a paternal sense to order, paternal is intelligence; religion is in any belief a deity or messiah saves you from what? Science on the other hand provides any interpretation to culture distinguishing relationships individual and collective... There was/is no intelligent design there was no knowledge before human material culture; (IE Adam & Eve parable) written by a poet; they became aware in vanity. The cause to philosophy is defining Cultural Anthropology a metaphysical dialog to the underlying casualness to mind in culture which conditions conscious objectivity. The bible is incomprehensible without cultural context.

At birth consciousness is a paradigm conditioned to physical awareness in physical attributes to dexterity subject to perceptive instinct in survival of environment. Interdependence conditions a paternal domestication in dependency unlike other species where consensual age is self-reliance in natural instincts verses the human development conditioned to the artificial intelligence of material posterity.

Drugs do not produce enlightenment... God is knowledge "Knowledge is cultural interpretation which is what consciousness is subjected to from birth... How do children develop language if not though observation in construal in perception to environment in adaptation to an existential material culture which conditions conscious interdependence in existential dualism... Adam & Eve became aware is its expression in the Old Testament, There is no physical collective consciousness only how any mind is impressioned by experience where there is often no reflection or mediation to many subjectivity's conditions thought in speculation challenging intuit to reasoning relativity lends to any confusions... Confucius says what you know you do not know how to describe invents gods of an existential authority postulating anyone knows one's own perceptions to existence...

Psychology is what conditions consciousness is culture in adaptation to material domestication IE Adam & Eve metaphors existential dualism seeking knowledge postulating a collective consciousness in vanity wanting approval...

Genesis Chap III

6-And when the woman saw that the tree was good for food, and that it was pleasant to the eyes and a tree to be desired to make one wise, she took from the fruit thereof, and did eat and gave also unto her husband with her and he did eat. 7 And the eyes of them both were opened, and they knew that they were naked; and they sewed fig leaves together, and made themselves aprons.

Physical awareness to material culture conditions human vanity

Culture conditions conscious objectivity "awareness" measuring countenance conditions existential polarities

Material domestication conditions procreation in codependence

Feminine vulnerability postulates subservience to provider in material domestications heredity

Is the following; a philosophic gathering, twelve step meeting or a spiritual revival

I Corinthians XIV

What is the topic? Watch you tube videos and "read the comments" on quantum physics and consciousness to know the parallel which has always existed especially during times of economic uncertainty.

Paradigm Shifts in conscious perspective are often triggered by dramatic changes or trepidations to environmental conditions is expressed in I Corinthians XIV where people try to describe their conscious experience also links to the new testaments expression of a period in despotism, based upon the postulated fall of the Roman Empire which is best compared to the 1930's and 1960's depressions and intellectual revivals. There is no collective conscious state of mind there is no one mind any individual state has an economic condition. No one knows what anything is or was, the only collective resolves are media impersonations' there is no true world view only what you know,

Paul; I Corinthians Chap XIV

Follow after charity, and desire spiritual gifts, but rather ye may prophecy. 2 For he that speaketh in an unknown tongue, speaketh not unto men, but unto god; for no man understandeth him; howbeit, in the spirit he speaketh mysteries. 3 But he prophesieth; speak unto men to edification, and exhortation, and comfort. 4 He that speaketh in an unknown tongue edifieth himself: but he that prophesieth edifieth the church. 5 I would that ye all spake with tongues but rather that ye prophesied: for greater is he that prophesieth, than he that speaketh with tongues, except he interpret that the church may receive edifying. 6 Now, brethren, if I come unto you speaking with tongues, what shall I profit you, except I shall speak to you by revelation, or by knowledge, or by prophesying, or by doctrine? 7 And even things without life giving sound, whether pipe or harp, except they give a distinction in the || sounds, how shall it be known what is piped or harped? 8 For if the trumpet gives an uncertain sound, who shall prepare himself to the battle? 9 So likewise ye, except ye utter by the tongue words easy to be understood, how shall it be known what is spoken? For ye shall speak into the air 10 There are, it may be, so many kinds of voices in the world, and none of them is without signification. 11 Therefore, if I know not the meaning of the voice, I shall be unto him that speaketh, a barbarian, and he that speaketh shall be a barbarian unto me 12 Even so ye, forasmuch as ye are zealous of spiritual gifts, seek that ye may excel to the edifying of the church. 13 Wherefore, let him that speaketh in an unknown tongue, pray that he may interpret. 14 For if I pray in an unknown tongue, my spirit prayeth, but my understanding is unfruitful. 15 What is it then? I will pray with the spirit, and I will pray with the understanding also; I will sing with the spirit, and I will sing with the understanding also. 16 Else, when thou shalt bless with the spirit, how shall he that occupieth thee in a room of the unlearned say, Amen, at thy giving of thanks, seeing he understandeth not what thou sayest? 17 For thou verily givest thanks well, but the other is not edified. 18 I thank my god, I speak with tongues more than ye all: 19 Yet in the church I had rather speak five words with my understanding, that by my voice I might teach others also, than ten thousand words in an unknown tongue. 20 Brethren, be not children in understanding: howbeit, in malice be ye children, but in understanding be men. 21 In the law it is written, with men of other tongues and other lips will I speak unto this people; and yet for all that will they now hear me, saith the Lord. 22 Wherefore, tongues are for a sign, not to them that believe, but to them that believe not, but for them which believe. 23 If therefore, the whole church be come together into one place, and all speak with tongues, and there come in those that are unlearned, or unbelievers, will they not say ye are mad? 24 But if all prophesy, and there come in one that believeth not, or one unlearned he is convinced of all, he is judged of all 25 And thus are the secrets of his heart made manifest; and so, falling down on his face, he will worship god, and report that god is in you of a truth. 26 How is it then brethren when ye come together every one of you hath a psalm, hath a doctrine, hath a tongue, hath a revelation, hath an interpretation. Let all things be done unto edifying. 27 If any man speak in an unknown tongue, let it be by two, or at the most by three, and that by course: and let one interpret. 28 But if there be no interpreter let him keep silence in the church; and let him speak to himself and to god 29 Let the prophets speak two or three, and let the other judge. 30 If anything be revealed to another that sitteth by, the first hold his peace 31 For ye all may prophecy one by one, that all may learn, and all may be comforted. 32 And the spirits of the prophets are subject to the prophets 33 For god is not the author of confusion, *but of peace, as in all churches of the saints 34 Let your women keep silence in the churches; for it is not permitted of them to speak; but they are to be commanded under obedience as also saith the law 35 And if they will learn anything, let them ask their husbands at home, for it is a shame for women to speak in the church 36 What! came the word of god out from you? or came it unto you only? 37 if any man think * himself to be a prophet, or spiritual, let him acknowledge that the things that I write unto you are the commandments of the Lord 38 But if any man be ignorant, let him be ignorant. 39 Wherefore brethren, * covet to prophecy and forbid not to speak with tongues 40 Let all things be done decently, and in order.*

Social gatherings are a great way to meet and get to know people; it is difficult to say the least to describe human experience of the cultures perversions with any accuracy or breadth in common understanding. Conspiracy and end time prophecy has always been a prevalent nuisance this prophets prediction is December 31, 2315

Infers there was a problem? Sort of like watching the movie Pleasantville the books had no words what do you mean there is something outside Main Street, (like the Wizard of Oz) there has always been a puzzle in artists assuming such a thing as one mind or such a thing as a defining moment or social all there, how would anyone know' can anyone know another's perspective or life story. Interdependence conditions existential dualism to cultural domestication. How does anyone know how to describe any owns social infirmities is a paradoxical assumption. Women by domesticating designs are considered to be the paternal nurturer is opaque initiate's codependent relations postulating paternal intuit taming masculinity contests dominion. Self-inflicted; dejection is an ambiguous valuation to reciprocal approval. Spousal abuses condition oppressive environments is ambiguous in contests to paternal solicitations.

Social infirmities are confused with global conflicts to indigenous ethnicity, traditions and religious ideals to cause in any mergence to culture conditions bias world views though all human culture has a paternal nuance in transfer to law

II Corinthians Paul Chap XIII-11

Finally, brethren, farewell: be perfect, be of good comfort, be of one mind, live in peace; and the god of love and peace will be with you.

There is no collective conscious mind' no one died changing the culture one mind shares a common cause for peace among people living by principles of an amicable accord

Was Jesus Paul's alter-ego or his poetical side so who was Paul trying to do one mind which is the underlying theme to the bible, (this was only written in a book not real life) what does anyone know as social awareness? There is no one mind connection only nationalism and intrinsic art making any connections to common experiences. What interpretation to the bible is there makes sense anyone knows how to comprehend its intent relative to anything in social principality. One mind postulates a principled people as law and you only know those you know there is no collective mode unless you are in a group like AA meetings one mind is also being one's own *Self so.* People living in poverty are subjected to far more social antipathy than the average fraternities. National pride postulates association to successful fortifications, what do nations defend? Freedom and liberty, where are they' no one is free we are bound to a yoke and bondage in faith to impotent resolves. What are oaths sworn on bibles anyone knows? To be honest; infers distrust are you a patriot knowing any accord.

For the world 911 triggered a sense of perpetuity in vulnerability taunting any dignity to any autonomous sovereignty and also set forth an economic surge followed by recession affecting economies.

Existential dualism is metaphoric in the Old Testament; *Cain* in measure "wanting approval" from an external authority of expectation for being successful as an individual making a contribution to an interdependent economy attaches to a world view in reflection making a contribution. Did the author of the bible play all the roles it is not impossible he may have had some assistance. Keep in mind that when the bible was written there was such a thing as industry and economy Hollywood does not expression history very well in most cases, visuals surely do impression people's minds that there are experts to his story, all words derive from conscious objectives for any understanding.

What is a true world view' now try to describe anything you know anyone knows, *let them edify*, does anyone need an interpreter. If there is anything to know anyone would know anything the same' a prophet is a philosopher assuming to know what a future holds, the only true physics are what the culture does is consistent in what technologies evolve increasing and obsolescing economic positions, today due to media technologies there are many prophets asking the same question; what are we doing assuming a collective mind changes, there is no physical collective mind there is only what is characteristic to the cultures primal envelopment or embodiment and any common associations. A social mind condition is what government does enforcing protective authority infringing privacy induces insurrections sources intellectual rivals is patriotic. Surveillance society is by and large the economics to technology in profitability "is capitalism" in human transfer; the world is our posterious replication to provisional domesticated permanence. Government expansion is an inelegant discord postulating principles. A house built upon the sand verses a house built upon a rock is in the Sermon on the Mount. Now read its poetic interpretation to the cultural malady. Liberty/ freedom is being your own mind and who or what pushes your buttons where forage is conditioned to free trade capitalism in "agricultural despotism" much as is the allegory to the film "Robin Hood," the original 1930's version.

Social/economic subjugation congests comprehension to experiences limiting any reflection or edification especially when we are on a treadmill being consumed by what we do in survival when we become what we do, capitalism is all a process of labels, culturally we become our environment, there is nothing to believe there is only what you know and what you know is what you experience is divisional due to what economy provides within any environmental opportunity. If you know what you know you will always be consistent. People test our memory. The fact is that no one ever died for anyone changing the malady to our culture. Assuming a collective mind knows a world view is superfluous, life becomes more fun when one realizes there is no point to anything and you simply exist knowing there is no way to describe how' assuming any why. What assumes any notoriety is any assumed contribution postulating obligation. What are social principles? Movies are sugar pills; materialism is a compelled addiction with no fault in adaptation. The thing is we are always at the end of the world and yet there is no such thing; prophecy always arises from economies in recession or depression. It was Armageddon during WWII. Global and Civil wars have in the past been caused by overpopulation stimulating unemployment causing economic trepidations truly questions any role to democracy conditioned to free trade capitalism expanding positions where there is no government.

In school did you memorize the names to all the people in your class feeling a sense looking forward thinking there was such a thing as maturity, why did the class ahead of us cause pretention, *is there a place to be in being,* what contrives generational views how does age consign consent shouldering personal responsibility? It would seem that having a source of income is the most conditioning medium extension to forage transfer in material urbanity which does condition all being. Education impressions a collective view to know is very inhibiting where there is no philosophic education disseminating commonality expressing any sense to an interpretive cause. When the bible was written things were nothing as they are today, it would be hard to imagine brings us to the arts ambitions looking at the present assuming any pathology to what fashions what tomorrow affords as a society conditioned to the expectations of law. Reality is opaque living within our own personal islands in provisional sanctuary; materialism is the true psychological environment to conscious subjectivity in heredity. When life becomes rudimentary there is no appreciation to the simple things where we can have any influence tries any imagination.

Resistance is any restriction in being your own thing in expression developing any talent which takes hard work in perseverance is a stressing factor, developing any talent while in youth has significance when you realize that later in life you have little to no personal time while in youth there is a significant pull wanting to be a social being conditions time in imagination rehearsing roles. Who are our mentors; people who are successful.

Thank you very much for another commercial selling me something I don't require; everyone needs a job to sustain their existence economic positions are embryotic in nature where there was no inalienable rights, *non-transferrable* diminishes any theoretical constitutions to sovereign delegate doctrines.

2002 War on terrorism only made things worse... and it is true that Islamic Muslim is a more primitive culture Sharia law is an eye for an eye tooth for a tooth has no virtue nor does believing in Mohammad, Jesus Allah God or Abraham... There is only one god of knowledge and that is interpretation to culture providing confirmations to common subjectivities in paternal polarities, economic influences and infirmities...

Transgression wants to size you up... Cain told Abel... people who are successful often are faced with social antipathies; we all experience this in school IE Columbine and Sandy Hook...

Religious faith has no intuitive virtue.... "Kung Fu guards no fence"

Embarrassing moments what are they?

Presence is an alluring shadow assuming any character association

God is a third party threat and lever for some

Willard is there such a thing as a rat race?

Notes:

Section Three

Justice what is the right thing to do episode 1 The Moral;

~You tube video

"Philosophy estranges us from the familiar not by supplying new information but by inviting and provoking a new way of seeing but and here is the risk once the familiar turns strange it's never quite the same again. Self-knowledge is like lost innocence however unsettling you find it it can never be unthought or unknown"

~ Law school professor

Any law giving you any right infers that you are owned is a very simple thing to comprehend, education to national historic bias is emblematic, history lessons should not be graded or anything subjective the only masteries are reading writing arithmetic and any physical science. Where there is no interest there is no art to erudition, self-interest applies to what one chooses as a profession.

What is it to have a job yet have no home and nowhere to stay working for minimum wage, what does government actually do providing "social security?"

An innate sense in fear of existence can trigger psychosis in assumption there is anything to know' there is nothing to know we all know our culture and any corruptions, there is no secret'

Citizenship postulates historic idiosyncratic memberships

Government education is socialism

A mind to perfection is any art in practice

Natural selection in capitalism is opportunity

A consumed mind leaves little room for imagination

Speculation to future stations always stimulates doubt

Childhood innocence steadily yields to the pressures in education conforming time

What people do in social degeneracy is very dispiriting

Money makes us slaves to our means

Knowledge is the poetical undercurrent to our culture; we all have knowledge in experience to adaptation from birth where there is no intuitive guidance, (knowledge) in "Adams Transcendence" as a metaphor to parental shortcoming which often transfers to religious postulations to paternal deities.

Paul; Corinthians I Chap VIII 1. Now as touching things offered unto idols we know that we all have knowledge puffeth up but charity edifieth. 2 And if any man think that he knoweth nothing yet as he ought to know. 3 But if any man love god, the same is known of him. 4 As concerning therefore the eating of those things that are offered in sacrifice unto idols, we know that an idol is nothing in the world, and there is none other god but one. 5 For though there be that are called gods, whether in heaven or in earth; (as there be gods many ;)6 But to us there is but one god, the father, of whom all things, and we | | in him; and one lord Jesus Christ, by whom are all things, and we by him. 7 Howbeit, there is not in every man that knowledge: for some, with conscious of the idol, unto this hour eat it as a thing offered unto an idol; and their conscience, being weak is defiled. 8But meat commendeth us not to god: for neither, if we eat, | | are we the better; neither, if we eat not, | | are we the worse. 9 But take heed, lest by any means this | | liberty of yours becomes a stumbling block to them that are weak. 10 For if any man see thee, which hast knowledge, sit at meat in the idols temple, shall not be conscience of him which is weak, be emboldened to eat those things which are offered to idols; 11 And though thy knowledge, shall the weak, brother perish, for whom Christ died? 12 But when ye sin so against brethren, and wound their weak conscience, yet sin against Christ. 13 Wherefore, if meat make my brother to offend, I will eat no flesh while the world standeth, lest I make my brother to offend.

An idol, is a symbol, symbolism is unknown belief in faith and worship as intuit or intuitive virtues know relativities to culture,

The afore mention of Jesus Christ and God of knowledge and idols are contradictions in the assumption there is an all seeing deity of knowledge, there is no collective consciousness god is knowledge to culture

---

as consciousness is conditioned to adaptation to material domestication is a bondage in the perpetuity of technological evolution providing positions in forage occupation from the Garden of Eden living naked in the wild. Not knowing the culture conditions memorialization's.

~ 1 & 2 has similar connotation to the professor's comment in self-knowledge twisted with Jesus innuendo, liberty is One Mind immunity no one can do for you, you are the only knowledge of your own life experience.

You know what you know has common senses, there is no cause there is nothing to know as memorialization conditions views to uncertainty

There is no collective mind to knowledge; "an idol" god "assuming creator" is an existential apparition ...

It may be possible that Jesus Christ represents what any artist or philosopher experiences in the strains to philosophic hermeneutics the crown of thorns a bit martyring being that it is the message not the cross as if humans are born of virtue

Mark IV the claimed words of Jesus describing hermeneutics in parable/ metaphor' 22 for there is nothing hid, which shall not be manifested; neither was anything kept secret, that it should come abroad. 23 If any man have ears to hear, let him hear. 24 And he saith unto them, Take heed what you hear: with what measure ye mete, it shall be measured to you: and unto you that hear, shall more be given. 25 For he that hath, to him shall be given; and he that hath not, from him shall be taken even that which he hath. 26 And he said, so is the kingdom of God, as if *a man should cast seed into the ground; 27 and should sleep, and rise night and day, and the seed shall spring and grow up, he knoweth not how. 28 For * the earth bringeth forth fruit of herself; first the blade, then the ear, after that the full corn in the ear. 29 But when the fruit is || brought forth, immediately he putteth in the sickle, because the harvest is come.*

Philosophic Hermeneutics is the challenge to philosophy where it is difficult to say the least to mediate experiences having any "consequential" understanding or self-dialog while developing any nomenclature vocabularies. Mark IV has anything to do with comprehension to relations as paternal beings in philosophic lecture; the author of the bible trying to describe what it was like to comprehend experiences into language in self-knowledge. ***22** we all have knowledge by experience in adaptation to our cultures embodiment where reflection has no mediation often relative to incursive exchange.* The truth is that there is a paternal discord where anyone wants to be part to a province stimulates any pretense to superficial natures. Things we experience condition sub conscious where there is/was no resolve and where memory is conditioned to repetitive cultural nuance often causative to memorialization is congesting or consuming much as is the constant strains to monetary weights. Mark IV makes an impression there is something to know however there is nothing to know there is no purpose it's not what you know it is how you construe anything valid and philosophic hermeneutics is nothing more than any practice in comprehensions to common cultural influences at times misdiagnosed in contingency. So if Mark IV and Paul are doing the same thing what was this Jesus guy spouting poems and parables *"ye squint* *at gnats and swallow camels*" the author expressing a sense of humor. What is it to explore the minds subjectivity to many things of experience where there is no comprehension to any association to cause in retort? Gilbert's law summary is an intriguing gesture in any attempt to comprehend social transgression. Anyone walking around speaking as Jesus in Mark IV would seem a bit odd and eccentric it is also a bit suggestive there is anything to know assuming anyone knows anything another does not' bestowing knowledge of what is already known as reality has no scripture it is perception knowing entanglement. Cultural literacy is the poetic political undertones to human interaction in societal governance.

Cultural Hermeneutics is the method in Philosophy to intrinsic arts construing conscious observance in cultural characteristics. It is difficult to say the least picturing in mind "summoned encounters" while developing an understanding converting anything to linguistic composition. Prompted recall is a practiced art and does stimulate better memory; memory recall is conditioned by knowing what's going on with people places and things has more organized cognizance. Contextual appliques are the most essential components for any realizations.

Hans Gadamer; "*providing a common language*," sharing notes; Gadamer and Heidegger were friends both working the same aspiration for providing an interpretation of the world being that their lives as was Einstein's had been subjected to much change and conflict. Paternal senses instigate expectations for a perfect world.

"Wanting approval is ego always looking over the shoulder often conditioned to doubt"

It should be pointed out that the writings in Mark are the same linguistic character as well as the same underlying aim as Corinthians I VIII, previously mentioned; *Mark IV 22* "*For there is nothing hid which shall not be manifested; neither was anything kept secret, that it should come abroad,*" is poetic assuming that we are all subject to the same physics in culture which often has no clarity or reflection which philosophy and intrinsic arts provide. The essential theme to Paul's writings is One Mind though there is no collective mind there is only any interpretation to our culture. One Mind is mentioned at least once written in II Corinthians by Paul who wrote "ten" epistles which is distinguished as being written by several authors though Paul is written in first person. One mind simply assumes the application to social principles and being centered. There is no collective aura it is only happenstance where any opportunity arises to make any connection and then there is what you are with family and friends or occupations acquaintance which is superficial when you do not know people. Do presidents and politicians assume to know us as one mind? There is no collective aura only any owns social sense epitomizing any barrier. There is no knowledge of any god, god is knowledge there is no intuitive comprehension to anything of simply faith or belief. Paul's quote makes the statement; you do not know what you do not comprehend is the consuming factor; we have knowledge though rarely take time in reflective edification to make sense to much experience. It is the nonverbal aspects to social posture which impression many conformities to cultural influences is the art to philosophic statements. Philosophic Hermeneutics is a painstaking process in social-reflection, Heidegger; "a long period of seeing and thinking" is the philosophers consuming occupation. You have to be a good detective to be a metaphysical scientist knowing anything unchangeable.

The Epistles' Paul is closest to who wrote the bible; it is so simple the bible is poetic/allegoric and written by one person being that no one writes as they speak and no two people write in the same linguistic style, what he was attempting to describe is that for the most part we have no comprehension knowing how to describe our experience is inhibiting as knowledge is competitive until we realize that the world we live in is mostly superficial unless we are on an intuitive plane. Another clue which points to the bible being written by one person is that Paul's character was written in first person however the only authorship (which no one ever talks about) in the bible is that of Paul, Titus, One Simon etc. assuming Mark Mathew Luke and John existed knowing the same exact story which was how they were written, we assume what we know is known when we attempt to communicate anything. The thing is that most people do not know how to describe their experiences without being inhibited and people in general are always looking for answers to life's questions.

It should be noted that Plato wrote epistle's and mentions One Mind in Meno which is an assumed dialog with Socrates however staged as if two people could articulate such an elaborate conversation writing anything down where recording devises had not yet been invented. It is likely sharing notes the bibles author used Plato's works providing confirmations to common questions summoned in philosophy. It is further noted that Socrates was made to drink poison by the then rulers of his time and died; his crime corrupting the youth and not recognizing the Gods acknowledged by the state. *

Being intimate is what a woman wants isn't it without being possessed. And yet you want to be with only the one that fits for life is the reason that in youth sex is more sacred than we assume, making love to someone is emotional and wants to be reserved for the one you want to procreate with which is still a transfer from instinctive nature of living in the wild undomesticated to any obligation in domesticating expectation. So why are divorce rates so high? Wait for the right one and you will know it. There is nothing immoral to premarital sex it is not what you do it is how one feels about what they choose to practice. Women can be lured into fixing their companion is not a good idea as it postures "paternal supremacy" is maladroit works either way, since paternal is intuit; this has something to do with high divorce rates and spousal abuse lending to children who come from broken homes can and does contribute to the causes in drug use and over populated prisons. There is much trepidation when anyone is in a relationship wanting a departure bound to a sense in obligation. Codependence postulates responsibility to another's sanctity. Getting married assuming anyone is going to make you complete is superfluous. And of course every marriage has any flaws, no matter what life is challenging. Communicating emotions is very intimidating and dejection is not a positive lure no one wants to be controlled. Ever heard of those having a midlife crises are the seeds to becoming a philosopher in reflection to experience.

I want you, The Band "Third Eye Blind"

Fix You, The Band "Coldplay"

Procreation is instinctive to any species survival

Early childhood is a mystic dream world

Any world view has a cultural sense

Taste gratification is the most common addiction

Reality has any intuitive connections in people you know

Deception has its own nature in pretention

Social studies postulates cultural characteristics

You can't fix people if you try you become a paternal puppet and initiate codependence if you argue much before marriage or feel seduced by dejection it's time to move on, being in love should sense a certain confidence to intuitive companionship being your own mind non possessive. Confidence has any attraction. Can we see the same malady as is the metaphor to the fable Adam & Eve so what is any true psychology to the environmental culture; are we not still subjected to the same material experience? Eve seduced Adam with the fruit of knowledge and what is knowledge but knowing what we all may experience. Virtue has any application to intuitive retorts in mind maintaining any sense to center having no lingering regrets in memorialization or the anchoring of the mind to anything past. Cain became consumed wanting approval from his own god sense wanting to be successful.

The world view is an economic genre

Bibles cast spells on people's minds in pretended knowledge

Quantum theory = cosmology and cultural anthropology

Obsession has many lures

Material possession conditions human procreation

Any oxymoron has a contradiction

A fear of what to know has an incriminating factor

Any chemicals affect physical endurance

Confucius say' Talking about people you don't know too well is non prudent

Following; the Koran has similarities to the bible, "they were sleeping" allegedly the author of the Koran was translating the underlying theme to the bible should be apparent by the mention of Jesus, "I am the nearest of all the people to (Jesus) the Son of Maryam, "Mary" "Interpret this parable to him so that he may understand it." People are not sleeping; we simply do not know how to describe our experiences is relative to the many layers to comprehension in adaptation to cultures social and economic epitome. Philosophy is self-education… Paul used the term edification or as the law school professor mentioned in his lecture, "self-knowledge is estranging" The reason bibles are so misinterpreted is that they are poetic parables or metaphor whereas metaphysics in philosophy provides scientific statements, metaphors are extremely difficult to decipher where there is no contextual prompt to experience. Contagious acceptances and pretended knowledge include the promotion to scriptures being translated as literal.

The Koran

Narrated Jabri bin Abdullah: Some angels came to the Prophet (Muhammad) while he was sleeping. Some of them said, "He is sleeping. Others said, His eyes are sleeping but his heart is awake," Then they said, "This is an example for this companion of yours." One of them said, "Then set forth an example for him." One of them said He is sleeping." Another said, "His eyes are sleeping but his heart is awake." Then they said, "His example is that of a man who built a house and then offered therein a banquet and sent an inviter (messenger) to invite the people. So whoever did not accept the inviter of the inviter, did not enter the house, nor did he eat of the banquet." Then the angels said. "Interpret this parable to him so that he may understand it." One of them said, "He is sleeping." The others said, "His eyes are sleeping but his heart is awake." And then they said, the house stands for paradise and the call maker is Muhammad and whoever obeys Muhammad, (*The mystery of God*) Obeys Allah; Muhammad separated the people (I.e.,) through his message, the good is distinguished from the bad, and the believers from the disbelievers)

God being intuitive virtue obeying one's own common senses

Narrated Abu Hurairah: Allah's Messenger said. "Both in this world and in the hereafter, I am the nearest of all the people to (Jesus). The son of Maryam (Mary) the prophets are paternal brothers; their mothers are different, but their religion is one

Prophet (Muhammad) b 570 ad

"*Their religion is one*" and what is religion but what culture does as scientific observance classifies culture and still no one knows any origins to life there is no why' why assumes intent' origins shall remain an eternal mystery, did life create life? If no one was told such a thing as a god existed they would never hold such a belief and then there are simply those who proscribe to contagious acceptances and religious assertions, an afterlife and religions and God as an authority postulates conceptual moral paradigms as any gate of entry is rather ambiguous where if such a place existed one would assume any relativity to knowing culture.

The bibles Cain metaphor; Cain left to the land of Nod, he fell to sleep,

Notes:

Section Four

In the following replace the word *Christ's* as *paternal beings* postulating knowledge as authoritarian and intuitive virtues'. Gradual>... transgression is what we experience in adaptation to an evolving culture having no instruction expected to fall in line in obedience to emblematic authorities knowing no goal. The universality in theories to a divine authority always originates within the concepts of creator or intelligent design assuming any purpose. There is no pre condition to human consciousness as culture is learned by adaptation repudiates any perceptions to saviors. No one ever died saving us from transgression as without transgression there would be no culture or conscious observations. Transgression is the learning curve where there is no comprehension. Lessor evolved indigenous tribes have limited developments to cultural linguistics does create any barriers to communications relative to religious fanaticisms though language is still universal to cultural relativity in patriarchal domestications; nonetheless simply we are all the same people on the same planet in the same universe from the same big bang chemistry and the same primal culture.

Paul II Corinthians Chap X, verse 7 Do ye look on things after the outward appearance? If any man trust to himself that he is Christ's let him of himself think this again, that, as he is Christ's, even so are we Christ's. 8 For though I should boast somewhat more of our authority, *which the lord hath given us for edification, and not for your destruction, I should not be ashamed; 9 that I may not seem as if I would terrify you by letters. 10 For his letters (say they) are weighty and powerful; but his bodily presence is weak, and his speech contemptible 11 Let such an one think this, that such as we are absent, such will we be also in deed when we are present, 12 For we dare not make ourselves of the number, or compare ourselves with some that commend themselves: but they, measuring themselves by themselves, and comparing themselves among themselves, \ \ are not wise.*

God is knowledge in culture, god is not the author of confusion… intuit knows what is does

Postulated paternal supremacy is the original cradle to authoritarian subjectivity, law imposes a moral conviction what is it? Is government your parent is a posterious pose and an intuit medium transfer as an interdependent culture. When one sense's a world changes it is not the world but one's own perception in any phase to individual objectivity applying to adaptation in sustenance procurement. Marriage conditions the view to an obligation sensing any expectation in dedication can limit relations. What is going on in general is the same primordial culture which has always existed conditioned to materials domestications. Law places constrictions on our minds conditions what to think is perversion questioning anything we consider doing lures an approving eye in suspicion postulating external authority. The force in what we do also conditioned by money is essentially a manner of material enslavement however' is cultural necessity in despotism to forage propentence.

Being in time is any seduction to being anything to anyone

Alcohol is a sedative which alters faculties exposes ones consumed thought and relieves pressures as a vehicle in escape much as does any mind altering substance. Chemical dependency can be progressive as a habitual mechanism. Alcohol lowers inhibitions is the reason people become addicted and may also lend to precarious occurrences in life which one may later regret. The power of suggestion is very influential regarding how much pretended knowledge can condition thought assuming that alcoholism is genetic or hereditary adds a sense in thinking oneself defective. The authoritarian; (*Your just like your father*) is very suggestive postulating any genetic condition. The cause to drug addiction is when one disconnects from the everyday pressures altering the mind where grandiosity has a sense of liberty where time seems irrelevant. However addiction is progressive concerning tolerance of the chemical combined with a sense of degradation being accumulative, one will chase that euphoric place in consciousness which can never be re known as it was the first time in the beginning to that journey which initiated the obsession. This is especially true with drugs such as opiates and cocaine. Marijuana is not deadly though can still be addictive in the sense to that particular state of mind which is different for each individual. When people get high they will often dream of being successful though more often will never follow through on their thoughts. Any addiction making a person a slave to chemical dependence is a very degrading aura. The obsession begins with an experimental mind or consumed mind. Seeking a mind in the state of bliss can be achieved another way called recovery knowing that the culture is what conditions our senses. And then there is what you have to do to sustain yourself in material comfort is a pragmatic goal.

Social economic value is a posterious reflection

Being at war changes the media presence on guard

One's own social condition is the people you know

Frustration is a consuming factor salivating tribulations

In promiscuity if you don't want to be stupid always use a condom

Visual images conjure thought in appearance to dislike and attraction

Language development practices intuitive comprehensions

If you don't know the culture you shouldn't be in government

Segregation is an economic condition and an interdependent nature

Hatred is often associated to irreconcilable adversity

Every economic genre has its own linguistic dialog

Children want a traditional family

Language nomenclatures cultural dispositions

Over reactive people try any patience

Socialist institutions fail safe programs

Sincerity is a noble cause

Religion is a defensive ideological partition as much is as law preserving any material maintenance is derogatory. You know what it is when you realize that time is when your conscious experience is subjected to an expectation to provide for yourself, now how do you describe the application to who you are being what you are to an interdependent culture in an economic transfer in forage propentence. Dualism is the existential experience the world is you in disguise. There are many facets to a physical world view; IE Quantum Theory and Quantum Mechanics have been misinterpreted by theoretical science if we do not know what our culture is then we have no clue what we are doing as one thing. Now what does anyone know since there is no directive nor any point to anything we simply evolve as an embryotic technological apparatus stimulating anything new to artificial intelligence. You do not know how other indigenous cultures are perceiving the world as technology brings people closer together as one thing, as human beings who are innovative because we have been born with certain physical attributes combined with the cognizant ability to describe experience which ambitions consciousness. You know physical medium extensions when you purchase a new car and it eventually becomes a part of you a technological accessory in the dualism to necessity.

Cultural interpretation has pertinence regarding how to apply any mind is real and something we have all gone through and never took a second glance assuming someone else would figure it out for us which is true regarding artists and philosophers providing common language for anyone to survey. The best way to communicate with anyone is best posed as a question, keeping in mind there is no collective mind knowing anything is the illusion to dual polarity in measure which is metaphoric in the bible. There is no all seeing eye in the sky there is no one keeping track of everything there is no true world view knowing casualness we are nothing but a distorted culture doing what we do to survive is the best way to describe it. There is no shame there is no sin there is only what we are subjected to in the partitions to economic attitudes and how we address our own conscience in the wakes of social hostilities. Capitalism is that impersonal because it is difficult to say the least that pride is condensed to conformity in expectation rather than the art to being what you are as one thing being inventive and still there is no collective mind knowing any cause. You know anything did not make sense when you have second thoughts; retraction in thought is an indication of one's intuition has many entwines.

There is no comprehension where there is no contextual point, the means functions are artificial intelligence AI, and cultural memorabilia is any celebratory fashions in adornment to envied posterity

If you know what you want to do for a living in high school you will learn as much out of school as you can on your own while working with mentors; there are fundamentals to any specialization

If you want to be smart you will not believe anything you do not comprehend including religious assertions

Authoritarian perversion wants to control every procedure to make sure everything and everyone is accounted for posturing recognition

Law assumes there is anything to know making a connection that we are one thing together having any purpose in being successful

Intelligent design; there is no collective conscious mind' god is a posterious apparition

Social entanglement is a kaleidoscopic speculation

There should absolutely be no pledge to allegiance in school (conformity to religion)

A positive mind links to physiological metabolism

Greed is a green eyed monster

There is much pretended knowledge as was proscribed in other writings by Hans Gadamer

What you do in sanctity of personal domain is your own individual right to privacy

When you are in youth there is a certain trust your elders know something you don't though you know what you know people are competitive

Stress is the leading cause to addictions

You find a sense in freedom getting your driver's license

Freedom is when nothing owns you and then there is taxation is truly a necessity to get things done though should have limits

Taxation owns you; national patriotic freedom is symbolic

Experience is the essence to any knowledge and accumulative

Education impressions a collective mind shares a historic world portrait

Idolization to memorialist national emblems condition authoritarian acuity

Sharia law knows a primitive culture postulates material domesticating morality

You are your own judge and jury of your own peers

There is no one mind there is no collective conscious view

Quantum Physics; purpose assumes any expectation for being successful and how we are successful as a thing drifting in space in an infinite universe where there is no meaning. The voice in your mind is your conscience wanting life to make sense unknowing where anything goes. There is no world view we all know there is no physical recording there is no movie following any script the world is largely reactionary. There are so many people out there trying to be Neo it's amazing that anyone assumes they know anything going on, there is what you do being superficial and there is what you communicate when you communicate anything and yet there is no physical recording paying anything forward being you only know the people you know and there is no collective mind recording any knowledge. A collective mind is any owns truths…

I exist you are not me I am not you and yet we are the same experience nothing owns us only any expectations'

We conform to law postulating authority is conforming stigmatizing virtue and what is virtue but knowing what you do makes sense

First impressions are pertinent postulating any connection

If you want to be in sports you will not use alcohol or any sodium, both lessen endurance'

Fatigue is conditioned by auto immune chemical withdrawal

Prediction; there will be fewer volcanic eruptions as the earth cools

Who said eating three meals per day is healthy when you can get by on one; stress causes many addictions to foods for comfort, you are confident when you are disciplined to your own health's concerns

Drug addiction obsessions obsolesce management skill to material proficiency

Authoritarian permanent records are conforming postulating restitution suggesting social justice conditions cultural perfectionism

Reality assumes a motion picture to time there is no true world view

Food addiction is any addiction to taste impressioning memory perception

Having any felonies doesn't look good on a resume, there is nothing immoral using drugs they should expunge record convictions' authoritarianism isn't intuitive its shaming you naughty little children, and then there are some when they are high do things they do not always remember brought the law,

Law is essential if there is to be any deterrent to crime'

The perversions to government law and taxation lend to civil unrest as was the case when Frederick Bastiat in France 1848-1850 wrote the book "The Law" during a period of insurrection lending to revolution in a time of despotism

There is no constitution there is no intelligent design

Drug addiction is progressive and self-destructive

Consciousness anticipates involvement individually and as one thing as the same occurrence. The more you know the less you know without trying. The world view intensifies as we adapt to our material appendage. When we are children we do not know anything about time, time is what the means does to our minds in expectation pulling our weight to an interdependent culture. Existential Dualism is the assumption there is a common world view. A collective view is what government impressions in transient causes. When you are in the early phases to life there is less pressure and a world view is open-minded and then we become what we do assuming identity to any idiosyncratic attraction.

Using marijuana can be non-compulsive in recreational use also can increase focus in a creative state of mind

People who are addicts often make up stories always changing

Movies impression minds with sadistic features

Time is the constant pull in personal expectation subject to cultural interdependence

He didn't know she knew something was askew

Without clothing we lose something very distinctive about us

What is it to be centered; meditation has two kinds' reflection and silence like taking a vacation from all the pressures, it is anything unresolved which becomes consuming

Social involves postulate true motives is any chess match to mind knowing who you are to anyone

You do not know anything anyone tells you concerning any others experience or motives

Prayer has no intuitive consult

While reading' another thought may appear' write it down

Marriage has a left temporal merger

Knowledge has any reflective directory changes nothing ever known

You have the right to remain silent there is no cause there is no law there is no authority only the law of the people

The main reason people use drugs is a fascination in getting high assuming enlightenment, though there is a plane of false liberty in untethered grandiosity short lived which begins any obsession chasing a place never to be re-experienced as the first time becoming a slave to addiction

Notes:

Section Five

"Immanuel Kant (1724-1804) agreed that the most essential metaphysical ideas—time, space, and causality—cannot be demonstrated. They therefore have to be accepted. But since there is no escape from them, they must Kant supposed, be the creations of the mind itself. The mind imposes these 'categories' on reality` because it cannot experience reality in any other way. According to Kant, it is pointless to try to prove or disprove such fundamental metaphysical categories; pure reason reaches its limits when it tries to depart what it is given in experience"

"Einstein approached this philosophical problem in a completely different way. While agreeing with Hume and Kant that we need metaphysics to make sense of reality as we experience it, Einstein pointed out that, from the standpoint of logic, all propositions are assumptions, "freely chosen posits," and this applies to those propositions closest to the experience of our senses, just as much as to the more 'abstract' propositions"

~Unknown author from the book seven geniuses in their own words

The following is a very significant statement pertaining to the context to quantum theory as to what is reality is the physics to cultural anthropology, the precept to Einstein can only be what are we doing and what drives our culture is the most difficult science to master Einstein lived through massive changes to technology which were accelerating innovations as electricity electronics mobility and communications were changing the economy while adding to educations specializations and also changing arms to war, it was and is assumed that war is the pathway to peace is obscure most physicists are theorists who become famous asking questions to what is existence.

Quantum Mechanics

"Within a theoretical system, said Einstein, a correct proposition is one which can be derived by logic from that system. A theoretical system as a whole can be provisionally accepted as true according to how well it is coordinated with the totality of our experience. Early in his career, Einstein had helped to found quantum theory, but he later became alarmed at the direction mainstream quantum mechanics was taking, under the influence of Niels Bohr and the Copenhagen Interpretation.' Einstein's famous remark that "God does not play dice with the Universe" expressed both his skepticism about the subjectivists and indeterminist aspects of quantum physics, and his ever present belief in intelligence behind the cosmos"

~Unknown author from the book seven geniuses in their own words

Intelligent design is evolving human innovation

"A categorical imperative would be one which represented an action as objectively necessary in itself, without reference to any other purpose."

"Fallacious and misleading arguments are most easily detected if set out in correct syllogistic form."

"Reason does not work instinctively, but requires trial, practice, and instruction in order to gradually progress from one level of insight to another."

"All our knowledge begins with the senses, proceeds then to the understanding, and ends with reason. There is nothing higher than reason."

"Seek not the favor of the multitude... But seek the testimony of the few; and number not the voices, but weigh them."

~Immanuel Kant

"In the sphere of thought, absurdity and perversity remain the masters of the world, and their dominion is suspended only for brief periods"

"As the biggest library if it is in disorder is not as useful as a small but well-arranged one, so you may accumulate a vast amount of knowledge but it will be of far less value than a much smaller amount if you have not thought it over for yourself."

"For an author to write as he speaks is just as reprehensible as the opposite fault, to speak as he writes; for this gives a pedantic effect to what he says, and at the same time makes him hardly intelligible."

"All truth passes through three stages. First, it is ridiculed. Second, it is violently opposed. Third, it is accepted as being self-evident."

"Every person takes the limits of their own field of vision for the limits of the world. It is only a man's own fundamental thoughts that have truth and life in them. For it is these that he really and completely understands. To read the thoughts of others is like taking the remains of someone else's meal, like putting on the discarded clothes of a stranger."

"National character is only another name for the particular form which the littleness, perversity and baseness of mankind take in every country. Every nation mocks at other nations, and all are right"

"Reading is equivalent to thinking with someone else's head instead of with one's own."

"We can come to look upon the deaths of our enemies with as much regret as we feel for those of our friends, namely, when we miss their existence as witnesses to our success."

"We forfeit three-quarters of ourselves in order to be like other people."

"Wealth is like sea-water; the more we drink, the thirstier we become; and the same is true of fame"

~Arthur Schopenhauer

You want to live in party mode where there are reliefs to the pressures of life's personal expectations and then there is the realization to your own responsibilities in self-preservation being attached to the material yoke so where is the balance when you want both to settle any discrepancies

A mind subject to force in expectation postulating benefit to inclusion in participation unknowing any plan summons anarchist views

What did you know you did not know to describe to anyone a collective mind knows anything you are involved in adopting any expectation having no interpretation

What is government verse's economy if economy is what conditions government is economy?

What the economy is to you as a child is very strange assuming a sustaining obligation as an institution

Punctuation precursor's comprehension

Notes:

Section Six

Consciousness has two virtues; the social being and the economic being having no decisive education. Media impressions a collective world view is superfluous suggestive and conforming postulating there is any governance knowing any design. A social view is a paternal association beginning in the home seeking intuitive commonality to cultural expectations for being successful. As children structure has significance priming any self-sufficiency coerces an authoritarian model in compliance to cultural interdependence in self-reliance in a paternal distortion to expectation. So what is consciousness if not analysis to experience? Now what are you doing looking for any position to simply afford any life style. What is subliminal and what is sub conscious is suggestive; subconscious is anything memorialized conditioning any view' you know what you know, what you do not know are any constitutes within any seduction.

We are born into an environment of visual and sensual stimulations having no intuitive knowledge to culture assimilating to domesticated expectations to be independent conditioned by an opaque authority providing any intuitive instruction to reason

Human culture did not begin with aliens…Infrastructure is embryotic to primitive ancestral transcendence

The question Why" is the first intuitive consultant

Reality is metaphysical existence to culture

How did the world communicate before media technologies

Reading stories adds paradigms to conscious experience

The bible limits philosophy in cultural anthropology as the Old Testament expressions cultural transcendence in the Adam Eve Cain parable, Cain sought approval from his god was denied became roth and left to the land of Nod. Cain and his plow metaphors agricultural despotism in cultivation in the use of tools evolving us from nomadic foragers establishing domesticating societal interdependence. There are multiple modes to mind as life is by and large subjective as the culture conditions expectation in pulling one's own weight. Metaphysics are in the bible though allegoric metaphoric and poetic in the attempt to provide a pragmatic world view.

Reality is what the culture does; now describe it as if reality exists knowing what we are subjected to knowing how to respond to the underlying divergence there is no visual picture only what you know and what media selects postulating politically correct

You do not know anything you were told in scripted memory' when you know anything you know what you know

Reality is the malady to your own realms exposure

Innuendo is a consuming influence knowing what you don't know

Religion is a parental crutch philosophy is religion knowing what you do

A common world view is generations changing fashions

Consciousness involves what you are a part to cosmic, social and economic

Subconscious is conditioned by any indoctrination to culture

Subconscious memory is largely conditioned by things having no retort or resolve

Subconscious is cultures programming hard drive data

If you know what you know you won't ever be fooled *

"Every great and deep difficulty bears in itself its own solution. It forces us to change our thinking in order to find it"

"Every sentence I utter must be understood not as an affirmation, but as a question"

"How wonderful that we have met with a paradox now we have some hope of making progress"

"If quantum mechanics hasn't profoundly shocked you, you haven't understood it yet"

"It is wrong to think that the task of physics is to find out how nature is. Physics concerns what we say about nature"

"Never express yourself more clearly than you are able to think"

"We are all agreed that your theory is crazy. The question which divides us is whether it is crazy enough to have a chance of being correct. My own feeling is that it is not crazy enough"

"A physicist is just an atom's way of looking at itself"

"The very nature of the quantum theory... forces us to regard the space-time coordination and the claim of causality, the union of which characterizes the classical theories, as complementary but exclusive features of the description, symbolizing the idealization of observation and description, respectively."

~Niels Bohr

How can anyone describe a true world view? The task in philosophy provides any interpretation

There is no world view anyone knows the same only the same cultural subjectivities and selected historic impressions presented as significant

There are two parts to reality the fact that the universe exists and you exist and what we do as a culture conditioning consciousness

If you have unique features and look stupid to yourself you may want to become an actor or comedian

Nationalism is a mysterious aura there is no collective copacetic mode knowing how to describe what's going on' what is it

What shapes personal character if not any chameleon gestures noble inclinations

The nature to quantum theory regards what our culture does impressioning world views' the nature to reality concerns what we do technologically a form of stimulation to innovations, technology has any culmination in extensions

How do you visualize a geographic perspective? Perception has photographic qualities are there any borders on the map or did someone make them up being acquisitive or is it a collective gesture. Now visualize a planet in the midst of an infinite universe where there is neither a beginning nor any end where anywhere is a center

Any future is not what you dream and the past is always a place where you have been. When you are young you are seduced assuming there is any government taking care of you believing anyone knows what anything does. There is no social congruent mind no one knows what anyone follows. Quantum physics 101 we are all a part of something that we do not know how to describe being there is no definition to goal

*

"And isn't it a bad thing to be deceived about the truth, and a good thing to know what the truth is? For I assume that by knowing the truth you mean knowing things as they really are Then we shan't regard anyone as a lover of knowledge or wisdom who is fussy about what he studies…[there are] two kinds of things the nature of which it would be quite wonderful to grasp by means of a systematic art… the first consists in seeing together things that are scattered about everywhere and collecting them into one kind, so that by defining each thing we can make clear the subject of any instruction we wish to give… [The second], in turn, is to be able to cut up each kind according to its species along its natural joints, and to try not to splinter any part, as a bad butcher might do… Phaedrus, I myself am a lover of these divisions and collections, so that I may be able to think and to speak"

"All learning has an emotional base"

"You can discover more about a person in an hour of play than in a year of conversation"

"Science is nothing but perception"

"When the mind is thinking it is talking to itself"

"We do not learn; and what we call learning is only a process of recollection"

"One man cannot practice many arts with success"

"Do not train a child to learn by force or harshness; but direct them to it by what amuses their minds, so that you may be better able to discover with accuracy the peculiar bent of the genius of each"

"Knowledge which is acquired under compulsion has no hold on the mind. Therefore do not use compulsion, but let early education be a sort of amusement; you will then be better able to discover the child's natural bent"

"I would teach children music, physics, and philosophy; but most importantly music, for the patterns in music and all the arts are the keys to learning"

"The purpose of the city if for the good of the whole, not the happiness of one class, both wealth and poverty are harmful"

"All things will be produced in superior quantity and quality, and with greater ease, when each man works at a single occupation, in accordance with his natural gifts, and at the right moment, without meddling with anything else"

"Nothing can be more absurd than the practice that prevails in our country of men and women not following the same pursuits with all their strengths and with one mind, for thus, the state instead of being whole is reduced to half"

"We ought to esteem it of the greatest importance that the fictions which children first hear should be adapted in the most perfect manner to the promotion of virtue"

"Every heart sings a song, incomplete, until another heart whispers back. Those who wish to sing always find a song. At the touch of a lover, everyone becomes a poet"

"Hardly any human being is capable of pursuing two professions or two arts rightly"

"Human behavior flows from three main sources: desire, emotion, and knowledge"

"For good nurture and education implant good constitutions"

"Man—a being in search of meaning"

~Plato

Plato was postulating a one mind collective consciousness; as we become more interdependent if our intention is progress we require better constitutions which links minds to the true physics to our culture Meno was such an endeavor, what governs human existence in any appreciation to equality isn't how it was written but what he meant though how is there any governance over capitalism and what is freedom if not the choice in provisional partiality

There is no such thing as one mind there is no social conscious recording knowing any cause you can never trust what government says without knowing any facts as they truly are and then there is how we all are expected to fit into an economy assuming anyone has anything figured out

Solitude has any appreciation being your own mind knowing you are the same thing subject to the same culture

What does anyone know to describe when genre conditions any language

There is what you know and what you apply yourself to making ends meet

Social conscience postulates common senses religious contentions are very distorting

Is the world becoming more perverted or did some simply become more aware is truly puzzling

A collective mind associates technologies innovations postulating progress has any enticement to convenience

Time is any appointment

An aggressive aura is a sign of insecurity

Dejection instigates suspicion in adverse dominions

There is nothing to know the more you know the less you know you know what you know when you know anything

Taste is a very incriminating addiction

Faith is suppositional in any realm to permanence

Does the picture on your driver's license make you feel like a suspect in a line up

A world view is a sociopathic conundrum in the courses to merging cultures postulating collective resolves

A perfect order would assume there is a position for anyone though capitalism is individual selection

You knew One Mind when you were a child before there was any assignment to idiosyncratic attachments in any cultural expectation girded to a bridle

Knowledge is the epitome to culture

*

"There is an ideal of excellence for any particular craft or occupation; similarly there must be an excellent that we can achieve as human beings. That is, we can live our lives as a whole in such a way that they can be judged not just as excellent in this respect or in that occupation, but as excellent, period. Only when we develop our truly human capacities sufficiently to achieve this human excellent will we have lives blessed with happiness"

"All persons ought to endeavor to follow what is right and not what is established"

"Men create gods after their own image, not only with regard to their form, but with regard to their mode of life"

"Men acquire a particular quality by constantly acting a particular way... you become just by performing just actions, temperate by performing temperate actions, brave by performing brave actions"

"Moral excellence comes about as a result of habit. We become just by doing just acts, temperate by doing temperate acts, brave by doing brave acts"

"Poverty is the parent of revolution and crime"

"A friend is a second self, so that our consciousness of a friend's existence... makes us more fully conscious of our own existence"

"The investigation of the truth is in one way hard, in another easy. An indication of this is found in the fact that no one is able to attain the truth adequately, while, on the other hand, no one fails entirely, but everyone says something true about the nature of all things, and while individually they contribute little or nothing to the truth, by the union of all a considerable amount is amassed."

~Aristotle

The truth is that we simply exist and there is no collective mind knowing what we do together as one embryotic culture and the truth is that our minds are conditioned by the same culture striving for something we want as success and what is it; material or social wanting to be your own thing fitting into the same thing

*

"There is grandeur in this view of life, with its several powers, having been originally breathed into a few forms or into one; and that, whilst this planet has gone cycling on according to the fixed law of gravity, from so simple a beginning endless forms most beautiful and most wonderful have been, and are being, evolved."

"Ignorance more frequently begets confidence than does knowledge: it is those who know little, not those who know much, who so positively assert that this or that problem will never be solved by science."

"The mystery of the beginning of all things is insoluble by us; and I for one must be content to remain an agnostic."

~Charles Darwin

Does the heat from the sun cause a gravitational pull to the orbiting planets makes sense being there is no spatial encumbrance? Now create an experiment where there would be no gravity would be a difficult experiment to produce a perfect simulation to our solar system, there is no resistance in zero gravity.

*

"Simplicity is the ultimate sophistication"

"It had long since come to my attention that people of accomplishment rarely sat back and let things happen to them. They went out and happened to things."

Principles for the Development of a Complete Mind: Study the science of art. Study the art of science. Develop your senses—especially learn how to see. Realize that everything connects to everything else."

"Anyone who conducts an argument by appealing to authority is not using his intelligence; he is just using his memory"

~Leonardo Davinci

Notes:

Section Seven

Ethics and morality concern virtue or intuitive self-consultation being conscientious knowing how to respond to subjective experiences or rather others emotional responses to what goes on in their lives without making anyone feel degraded knowing that life is emotional when things simply have not worked out in our dream worlds for a perfect life. There is no perfection being there is no true cause to which we apply ourselves working together assuming a purpose. There is no purpose only technologies evolution to innovations conveniences. There is too much pressure wanting to be successful and what is success? If we want a more perfect world we would adopt healthier economic values while providing for those who do not know how to provide for themselves disregarding government debt which has been influenced due to the discourse of ignorance unknowing cultural probabilities. What is common wealth if such a thing actually exists projecting anyone must pull their own weight as a slave in bondage to a material culture as technology conditions economic positions keeping everything glued together to our interdependence? One mind is a posterious assumption there is any world view taking notes assuming a social mind knowing what we do as one culture. A collective social aura is your own disposition having any sanctuary.

This writing describes consciousness as well as nutrition and auto immunity as the body has its own repair mechanisms which can and are affected by certain carcinogens or *anti bodies* which may contribute in the causes to cancer. Eat healthy and you will be alright. Alcohol for instance burns tissue which activates auto immunity causing stress during withdrawals, it should also be mentioned that sodium is a drug and has no nutritional value and is only used to enhance flavor contributing to food addictions. Taste impressions memory perception it is also true for anyone that sodium lessens endurance conditioning metabolism contributing to obesity & poor health. Metabolism is also affected by what we ingest having no molecular residency, onions are carcinogens so are seeds, flour for instance is a seed not a fruit. The best nutrients are greens which also speed up digestion. Roots such as onions and potatoes are not fruits and have little to no nutritional value. Non nutritional substances convert to cellulose: how does alcohol cause fatty tissue? There is no nutritional component. Certain foods take longer to digest which also causes stress. Meat and flour do not digest well congesting the digestive tract. Stress is the number one cause to addictive tendencies. There is much pretended knowledge in the world especially what (media says) being that most doctors and physicists do not know all physiological occurrences clinical studies are often false and misleading which spread as being factual, one can only know or sense how anything affects them.

Considering mortality there is no such thing as knowing anything you simply end up cosmic dust so what is the concern though the miracle of life makes no sense it ends postulates creators and afterlife's

The world view for most people is a congestive fog

You can change others attitudes not buying in to negativity

Sympathy cards initialize codependent paternal patrons

If Helen Keller could get over being deaf and blind what is the trial

Education'; weights and measures what are they if not components to formulas

High School is the most socially perverting experience contesting image

Forage propentence; there's no one there to do anything for you

Reality is what you know when you have no idea how to apply yourself to anything is mystical unknowing what any future affords

Reality is epitome to culture

Based upon laws moral conviction Jerry Springer shows should be outlawed

If you want a good pizza you will mix a little love and company into the recipe everyone wants to have a specialty

Media stimulation is addictive for many feeling connected to the world

Network corporate media is a narrow collective perspective

Expertise to anything is trial and error provides experience to know the difference

There is no problem you simply exist and do not know there is any why

Economic positions divide people even though nothing works if nothing works together is the interdependence which invented democracy

There is a time for work and there is a time for music taking a vacation in melancholy reflection

There is no true world view only what postulates significance becomes adaptable assuming a collective mind in measure to progress

Martyring moments in thought conjure reason

If you live in a small town with nowhere to go there is more cause to addiction

Salome moments are when you realized nothing makes sense there is anyone knows how to keep track of the world though we must remain regimented, there is the pressure to time and reflection when you are flowing with the arts making any connection

Why can't life just simply be a good time for everyone without any competition and still there is no social gathering other than the arts and then there are religions, sports and the circus walking tight ropes challenging any expertise being daring proving conviction and confidence.

WARNING; THE FOLLOWING CONTENT MAY BE DISTURBING HOWEVER TRUE...

Philosophy 101 there is no directive there is no collective social mind there are only cultural landscapes; there is no authority being there is no interpretative cause only impressions of the arts, innovations in technologies and medicine. Law infers something owns us; freedom and liberty are "*symbolic constants*" law providing any right is totalitarian servitude. Law infers intent there are no laws to the universe there are only principles to how anything functions, authoritarian oppression is a posterious warden. Law imposes a moral conviction there is no separation between church and state, authoritarianism is religion "forcing" a moral standard under a postulated god authority. There is no instruction manual defining what we do as one culture only the freedom to religious consult to imaginary people seeking answers to what resolves? Economic transgression in forage propentence transfer is a cavernous perversion relative to how education instills trust in an illusory national guardian.

Criminal is the intent to harm physically or mentally another person including property (physical mediums) all mediums have any emotional attachment to existential polarities. Using drugs is not criminal drugs are used to escape life's pressures, marijuana is a plant and has medicinal qualities and does also enhance focus as it motivates awareness and should be legal anywhere. What science is there in opinions to criminal when there is no definition to intuitive competence? Ignorance is criminal assuming any authoritative position. Is competence considered authoritarian obedience? Comprehension has shortcomings relative to any non-intuitive compliance postulating instructive causes. What defines intelligence should be considered common sense. Authoritarianism is an oppressive agent causes a paranoiac stigma defying kosher fraternity. Is government your parent' being told how to think ostracizes intuitive virtue reasoning one's own convictions. "The law" in extension of aggression is a vengeful appliance in the intent to punish or harm is also transgression and criminal. There is no collective social mind there is no restitution to judiciary vengeance ; a common view is societal corruption sighting aggressive behavior. An education impression's there is a common world view collective mind following a story impressions national characterization; "*conformed bias*" educational settings also segregate posterious social positions. Capitalism is corruption in a game of monopoly, the truth known there has been no education preparing anyone for adaptation to such a distorted culture. Religious faith and worship is an exploitive agency.

Police state is a transient cause and catalyst to revolutions restoring liberty' *"one is one's own domain"* *"Intelligence is paternal says the law"* Communism is an authoritarian corruption in the derelictions to emblematic constitutions'

Liberation is a penitent noise in the confines to an illusory deliberation of a posterious collective mind

*

Alcohol addiction conditions auto immunity constantly in repair which can be remedied by drinking pure water having healing properties. The body is over 90% water: water can be used to flush out any chemical. No matter what anyone tells us salt has no nutritional value it is not a nutritional component and does burn tissue. Eat some potato chips; what happens to the tongue it burns and absorbs salt immediately and causes stress to the heart. The heart is a muscle being how exercise is significant even simply taking walks on a regular basis. Blood pressure is affected two ways; too much social and economic pressure initiating anxiety and what one ingests causing any physical rejection. Most additive tendencies are induced by consuming factors while wanting to escape. Energy drinks condition auto immunity causing withdrawal triggering any compulsion to feed addiction, anything which induces artificial energy requires refueling is a degrading experience in being chemically dependent.

Scent and taste were once used as forage beings as indications to carcinogens anything that bites would be considered an anti-body not being molecular organic nutritional replacement

Children have an innate sense of fear bleeding not knowing how white blood cells form platelets in the body's own powers of healing.

Antibodies cause any reaction over working auto immunity may link to cancer and degenerative aging

We as humans are addicted to stimulation physical social and emotional "a cultural malady" which the bible does epitomize, especially Paul's writing describing his life being the likeliest history in its author trying to distribute "*his knowledge*" through the story of Jesus. The bible can be read as a metaphoric guide in confirmations to personal knowledge though is too misleading in the story of Jesus as a savior. No one can change the life you have lived no one knows your own reality tunnel or world sense no one knows who you know no one knows how you know anyone. The author wrote the Old Testament first which was any attempt to quantum theory providing an allegoric genealogical account to material cultural envelopment, "*the preceptual context to reading the bible*", being nothing written is literal as there was no one there to write everything down from any beginning; Paper did not exist as what drives our culture is technologies innovations providing efficiencies and conveniences in any origins to economic interdependence. Cultural Anthropology can only perceive what does exist tracing anything backwards, what does exist can be known (*deduced or construed*) all else speculation within what anyone is led to believe assuming any historic authorities. The artist's enigma; there is no one responsible for human existence you know there is no such thing as one mind when you walk through the shopping mall. Social principality is inelegant regarding how economy segregates presumptive classifications. Postulations in god creators assume a deity knows ones thoughts and mind is any primacy to existential dualism IE the Cain metaphor assuming a purpose to measure has an accumulative affect in memorialization. (*Marshal McLuhan "consumer becomes consumed"*)

The underlying ideological debate reasons there is "*no interpretation*" to what we do as one culture delegate's postulations to authorities. Scripted religious ascertations cannot be dialoged being there is no knowing anything written in books, the bible was written by one person though claims translations from Hebrew and Greek, however the linguistic dialect (Thou) is nostalgic English, and people then still spoke plain language not as the theaters stage. No two people write the same style nor is it possible for anyone to keep track of any period of time following what's going on then writing down "*knowing dialog*" anything said points to the fact that nothing in the bible could possibly be authenticated, Paul's story "*may be*" the true authors life however he never knew a person named Jesus may be defensive to some however what would condition any change to a social epitome assuming any magical powers of a postulated messiah. Reading the bible literally assuming forty people wrote it and knew the same stories seventy years later tells us something about theologians limitations in any interpretations to metaphor or allegoric stories. Does the Immaculate Conception and the birth of Jesus metaphor the ignominy of an unwed mother? Connecting to the Old Testament' the genealogy of Joseph is of Abraham its genealogy beginning with the metaphoric fable of Adam Eve Cain and Abel. The setting to the story of Mary and Joseph is Jerusalem however the testimonial accounts of Jesus by Mathew, Mark Luke and John are English names. And of course anyone still has the right to believe anything though no one today was there as witness. If Jesus was known at birth why did he show up so many years later wanting to be baptized if he was the *"Son of God"*? Children are impressioned by the calendar and the holidays, Christmas impressions Jesus as a historic figure though there is no common knowledge regarding anything intuitive to the written scriptures/teachings only that this person died saving people from sin is suggestive. Easter is subjective and martyrizing assuming Jesus died on a cross and still exists and will return and do what? The story of Jesus should be a comedy like Monty Pythons "Holy Grail", now follow the story and try to imagine some long haired hippy narrating a poetical social commentary *"The Sermon on the Mount"* from memory... It never happened it was only written in a book and it is also true the calendars were changed less than 2000 years ago as the bible was allegedly translated beginning in 1604 and never published till 1611 by King James. It is just true no one knew how to write anything down during times in despotisms reversal *IE 1930's and 1960's*. Economically the world would not function if there were multiple time lines nothing would connect. By genealogical accounts it is a fact that there is on record a change to the calendars in 1750. The fact there was no printing press until 1430 does suggest Medias of the past and keeping records would have been quite ambiguous feigning any accuracy to any historic account through bias emperor and monarchic doctrine. There were no dark ages between Roman civilizations alleged period, (BC) technologies evolving economies have always been consistent expanding material culture' though "primeval" cultures surely were more barbaric; technology has its own ancestry in consistent innovation, no one knew how to rewrite history too well so we are conditioned to what education impersonates as history postulating any historic paradigm has been recorded.

Memorizing anything assumes intuit, intuit is contextual comprehension

Maybe god was a chemist who mixed strange elements together which ended in disaster causing the big bang. Existence itself is very mind boggling and mysterious relative to the infinite depths of the universe and any and all potentials.

You tube—Monty Python "The Sermon on the Mount" Mathew V-VII

Transgression, memorialization, practice dogma/virtues

Mathew 5:13 ye are the salt of the earth: but if the salt have lost its savour, wherewith shall it be salted? It is therefore good for nothing but to be cast out, and trodden under foot of men. 14 ye are the light of the world. And a city that is set on an hill cannot be lit 15 Neither do men light a candle and put it under a bushel, but on a candle stick and it giveth light unto all that are in the house ;"Shares similarities to the aforementioned Koran"

5:29 And if thy right eye || offend thee, pluck it out, and cast it from thee: for it is profitable for thee that one of thy members should perish, and not that thy whole body should be cast into hell

5:30 and if thy right hand offends thee, cut if off and cast it from thee: for it is profitable for thee that one of thy members should perish, and not that thy whole body should be cast into hell

The last word wants a virtuous pardon in being right

You-Tube; Monty Python-Killer Bunny Holy Hand grenade

The following narrated in Monty Python dialect

The "lord's prayer" is part of the poem of the Sermon on the Mount; Chap 6:-9 "Our Father" is a mystic polarity and often used as prayer out of its poetical setting of social subjugation

6:12 and forgive us our debts, as we forgive our debtors 13 and lead us not into temptation; but deliver us from evil; for thine is the kingdom and the power and the glory, forever, Amen, 14 For if ye forgive men their trespasses your heavenly father will also forgive you; 15 But if ye forgive not men their trespasses, neither will your father forgive your trespasses, 16 moreover, when ye fast, be not as the hypocrites, of sad countenance: for they disfigure their faces, and they may appear unto men to fast, Verily I say unto you, they have their reward, Grasshopper, 17 But thou, when thou fastest, anoint thy head and wash thy face, 18 That thou appear not unto men to fast, but unto thy father which is in secret; and thy Father, which seeth in secret, shall reward thee openly 19 Lay not up for yourselves treasures upon earth, where moth and rust doth corrupt, and where thieves break through and steal 20 But lay up for yourselves treasures in heaven, where neither moth nor rust doth corrupt, and where thieves do not break through nor steal, 21For where your treasure is there will your heart be also 22 The light of the body is the eye; if therefore thine, thine eye be single, thy whole body shall be full of light 23 But if thine eye be evil thy whole body shall be full of darkness, if therefore, the light that is in thee be darkness, how great is that darkness. 24 No man can serve two masters; for either he will hate the one, and love the other: or else he will hold to one, and despise the other, ye cannot serve god in mammon.

6:34 take therefore no thought for the morrow for the morrow shall take thought for the things of itself, sufficient unto the day is the evil thereof

22-23 metaphor transgressions distortion to composure…

7:3 and why beholdest thou mote that is in thy bothers eye, but considerest not the beam that is thine own eye? 7-4 Or how wilt thou say to thy brother, let me pull out the mote out of thine eye; and, behold, a beam is in thine own eye? 7-5 Thou hypocrite! First cast out the beam out of thine own eye; and then thou shalt see clearly to cast out the mote out of thy brothers eye 7-6 Give not that which is holy unto the dogs; neither cast ye your pearls before swine, lest they trample them under their feet, and turn again and rend you 7-7 ask, and it shall be given you; seek and ye shall find; knock and it shall be opened unto you 7-8 for every one that asketh, receiveth, and he that seeketh findeth, and to him that knocketh it shall be opened

Open mindedness is next to godliness one mind is knowing what you know, negative dispositions seduce paternal sentiments challenging virtue; Transgressions are the seeds to futility

Third eye Blind> Jumper | | Incubus> Pardon Me'

Authoritarian paternal supremacy wants to fix you up is not who you know, no one knows another's perceptive virtues

Taunting has seductive lures to illicit behaviors to control through fear is defamatory

Back to the Future Shamus wants you to know something

Ego has a pendulum polarity with comparison

Inhibition is the leading cause to alcohol abuse

Law postures a social responsibility to be proper citizens

Postulating natural authority' there should be no laws condemning nor condoning homosexuality

There is no cause there is no authority only any knowledge to cultural casualness

Notes:

Section Eight

A world view is what goes on with family, what goes on in community, what goes on within any industry supporting us, what goes on in your state, what goes on to a national level and what goes on globally as one culture conditioned to technologies dispersing economy in agricultural partition to forage propentence. And then there is what any future dreams or remembers of a past assuming a collective mind recording any relevance.

Who has time to think and what is being in thinking? "And then there is how we all are expected to fit into an economy assuming government has anything figured out? There is what you know and what you apply yourself to, to make ends meet. There is no such thing as one mind there is no social consciousness being recorded knowing any cause you can never trust what government says without knowing any facts as they truly are"

What is reality is there such a thing as a social mind following anything? There is nothing to know it is how you know anything makes any sense, the means functions are all artificial intelligence. Is there anything to know? There is no cause there is nothing to know it is not what you know it is how you know anything. There is no cultural connection working together so how is there any union? There is no motion picture in time only what anyone knows though whose perspective assumes a world portrait or physical film recording everything, IE in part the idea to an all seeing eye in the duality of interdependence. .

*

"The medium is the message. This is merely to say that the personal and social consequences of any medium—that is, of any extension of ourselves—result from the new scale that is introduced into our affairs by each extension of ourselves or by any new technology"

"Societies have always been shaped more by the nature of the media by which men communicate than by the content of the communication"

"Ideally, advertising aims at the goal of a programmed harmony among all human impulses and aspirations and endeavors. Using handicraft methods, it stretches out toward the ultimate electronic goal of a collective consciousness"

"One of the effects of living with electronic information is that we live habitually in a state of information overload. There's always more than you can cope with"

"As technology advances, it reverses the characteristics of every situation again and again. The age of automation is going to be the age of 'do it yourself'"

"As the unity of the modern world becomes increasingly a technological rather than a social affair, the techniques of the arts provide the most valuable means of insight into the real direction of our own collective purposes"

"A commercial society whose members are essentially ascetic and indifferent in social ritual has to be provided with blueprints and specifications for evoking the right tone for every occasion"

"A point of view can be a dangerous luxury when substituted for insight and understanding"

"Everybody experiences far more than he understands. Yet it is experience, rather than understanding, that influences behavior"

"Innumerable confusions and a feeling of despair invariably emerge in periods of great technological and cultural transition"

"Our Age of Anxiety is, in great part, the result of trying to do today's job with yesterday's tools and yesterday's concepts"

"The spoken word was the first technology by which man was able to let go of his environment in order to grasp it in a new way"

"We shape our tools and afterwards our tools shape us"

"Art at its most significance is a Distant Early Warning System that can always be relied on to tell the old culture what is beginning to happen to it"

"We become what we behold we shape our tools and then our tools shape us"

"We drive into the future using our rear view mirror"

~Marshall McLuhan

*

"A creative man is motivated by the desire to achieve, not by the desire to beat others"

"To achieve, you need thought. You have to know what you are doing and that's real power"

"You can avoid reality, but you cannot avoid the consequences of avoiding reality."

"A desire presupposes the possibility of action to achieve it; action presupposes a goal which is worth achieving"

"Do not ever say that the desire to "do good" by force is a good motive. Neither power-lust nor stupidities are good motives"

"Civilization is the progress toward a society of privacy. The savage's whole existence is public, ruled by the laws of his tribe. Civilization is the process of setting man free from men"

"Achieving life is not the equivalent of avoiding death"

"Every man builds his world in his own image. He has the power to choose, but no power to escape the necessity of choice"

"Just as man can't exist without his body, so no rights can exist without the right to translate one's rights into reality, to think, to work and keep the results, which means: the right of property"

"Potentially, a government is the most dangerous threat to man's rights: it holds a legal monopoly on the use of physical force against legally disarmed victims"

"The only power any government has is the power to crack down on criminals. Well, when there aren't enough criminals, one makes them. One declares so many things to be a crime that it becomes impossible for men to live without breaking laws"

"A "collective" mind does not exist. It is merely the sum of endless numbers of individual minds. If we have an endless number of individual minds who are weak, meek, submissive and impotent—who renounce their creative supremacy for the sake of the "whole" and accept humbly that the "whole's" verdict—we don't get a collective super-brain. We get only the weak, meek, submissive and impotent collective mind."

~Ayn Rand (1905-1982) Russia

*

"People don't like to think, if one thinks, one must reach conclusions. Conclusions are not always pleasant."

"Toleration is the greatest gift of the mind; it requires the same effort of the brain that it takes to balance oneself on a bicycle."

"True happiness is not attained through self-gratification but through fidelity to a worthy purpose"

"You don't love someone for their looks, or their clothes, or for their fancy car, but because they sing a song only you can hear."

"We can do anything we want as long as we stick to it long enough."

"Change: A bend in the road is not the end of the road… Unless you fail to make the turn."

"To keep our faces toward change, and behave like free spirits in the presence of fate, is strength undefeatable."

"Self-pity is our worst enemy and if we yield to it, we can never do anything good in the world."

"Knowledge is love and light and vision."

"Relationships are like Rome—difficult to start out, incredible during the prosperity of the 'golden age', and unbearable during the fall. Then, a new kingdom will come along and the whole process will repeat itself until you come across a kingdom like Egypt... that thrives, and continues to flourish. This kingdom will become your best friend, your soul mate, and your love."

"So much has been given to me I have not time to ponder over that which has been denied."

"Four things to learn in life: To think clearly without hurry or confusion; To love everybody sincerely; To act in everything with the highest motives."

"Literature is my Utopia. Here I am not disenfranchised. No barrier of the senses shuts me out from the sweet, gracious discourses of my book friends. They talk to me without embarrassment or awkwardness."

"There is beauty in everything, even in silence and darkness."

"Blindness separates people from things; deafness separates people from people."

"There is joy in self-forgetfulness. So I try to make the light in others' eyes my sun, the music in others' ears my symphony, the smile on others' lips my happiness."

"I long to accomplish a great and noble task, but it is my chief duty to accomplish small tasks as if they were great and noble. The world is moved along, not only by the mighty shoves of its heroes, but also the aggregate of the tiny pushes of each honest worker."

"I who am blind can give one hint to those who see: Use your eyes as if tomorrow you would be stricken blind. And the same method can be applied to the other senses. Hear the music of voices, the song of a bird, the mighty strains of an orchestra, as if you would be stricken deaf tomorrow. Touch each object as if tomorrow your tactile sense would fail. Smell the perfume of flowers, taste with relish each morsel, as if tomorrow you could never smell and taste again. Make the most of every sense; glory in the beauty which the world in all the facets of pleasure reveals to you through the several means of contact which Nature provides. But of all the senses, I am sure that sight is the most delightful."

~Helen Keller

*

"We've all heard that we have to learn from our mistakes, but I think it's more important to learn from successes. If you learn only from your mistakes, you are inclined to learn only errors"

"When every physical and mental resource is focused, one's power to solve a problem multiplies tremendously"

"Action is a great restorer and builder of confidence. Inaction is not only the result, but the cause, of fear. Perhaps the action you take will be successful; perhaps different action or adjustments will have to follow. But any action is better than no action at all"

"Formulate and stamp indelibly on your mind a mental picture of yourself as succeeding. Hold this picture tenaciously. Never permit it to fade. Your mind will seek to develop the picture... Do not build up obstacles in your imagination"

"Our problem is to become acquainted with our own selves, letting our personalities loose upon the world for the sheer adventure of their full development and in the positive hope that they may in their own way lift the level of humanity"

"One of the greatest moments in anybody's developing experience is when he no longer tries to hide from himself but determines to get acquainted with himself as he really is"

"Plant seeds of expectation in your mind; cultivate thoughts that anticipate achievement. Believe in yourself as being capable of overcoming all obstacles and weaknesses"

"Drop the idea that you are Atlas carrying the world on your shoulders. The world would go on even without you. Don't take yourself so seriously"

"If you paint in your mind a picture of bright and happy expectations, you put yourself into a condition conducive to your goal"

"Stand up to your obstacles and do something about them. You will find that they haven't half the strength you think they have"

"Every problem has in it the seeds of its own solution. If you don't have any problems, you don't get any seeds"

~Norman Vincent Peale

Being constructive if you enjoy what you are doing your mind will not be complacent

If you have any metal products to discard place them by the street for the scrappers

Wanting to become an actor in Hollywood has an inhibiting factor

Warning edible THC products have a creeping affect use sparingly

An eye for an eye a tooth for a tooth represents transgression and is Sharia Law

Middle Eastern ancient feuds have any territorial claims

For maximum duration masturbate thirty to sixty minutes before intercourse

Reality is a cultural condition in perception to time in mystical frolic

Children are naïve and impressionable as perception to culture is not instinctive

Cain slew Abel metaphors transgression in comparison in a competitive nature

Government conditions an existential authority in democratic transfer postulating a constitution

No one can change what another has experienced there are no saviors

Acting stupid to fit in curbs scruples

Money conditions all aspects to life

Deviation has a certain way in social resentment

There is no matrix code one mind there is no collective mind resolve' knowledge edifies individual perception

Society is a class menagerie

Marriage has many expectations in spontaneity

Memorization does not focus on comprehension

Equality is a classless society

Notes:

Section Nine

EPISTLE TO THE AUTHORITARIANS BY FRED,

Paul Romans Chap I, *19 Because that which may be known of God, is manifest || in them: for God hath shewed it unto them. 20 For the invisible things of him from the creation of the world are clearly seen, being understood by the things that are made, even his eternal power and Godhead; || so that they are without excuse: 21 Because they knew not God, they glorified him not as God, neither were thankful; but became vain in their imaginations, and their foolish heart was darkened, 22Professing themselves to be wise they became fools; 23 and changed the glory of the uncorruptible god into an image made like to corruptible man, and to birds and four footed beasts, and creepy things. 24 Wherefore God also gave them up to uncleanliness through the lusts of their own hearts, to dishonor their own bodies between themselves 25 Who changed the truth of God into a lie, and worshipped and served the creature more than the creator who is blessed forever. Amen 26 For this cause God gave them up unto vile afflictions for their women did change the natural use into that which is against nature 27; And likewise also the men with men leaving the natural use of the women, burned in their lust one toward another; men with men working that which is unseemingly, and receiving in themselves that recompense of their error which was meet*

"Because that which may be known of God is manifest in them" God is knowledge in culture we all have knowledge by experience in adaptation to material domestication in expectation

to be successful in virtue... This is all philosophy God is knowledge and the only thing common in metaphysical science... knowing cause... Without excuse we all know right from wrong intelligence is paternal... The basis to law is a paternal sense to order. Criminal is any intent to harm another individual... motive is conditioned by malevolence or revenge

Cultural literacy; culture conditions conscious objectivity in adaptation to material domestication; There is no collective conscious mind there are only individual perspectives to environmental influence conditioned by our material culture.

Perception of culture changes a view terming contributions

Hermeneutics distinguishing any cultural nuance defines language in conscious objective in cognition to myriads of encounter. Pretended knowledge apparitions external deity sources; we all have knowledge by experience to cultural adaptation in domestications and infirmities; Godhead is another word for "*intuitive natures*" knowing any similarity in social reserves to virtuous standards. It can be difficult being judicious in retort when others expect anything from us as wanting to help often involves interventions is invasive assuming authority. Being virtuous also applies within how to deflect social taunts without being seduced by retaliatory causes; being without excuse knows people share common senses to illicit behaviors where often demeanor is conditioned by any environmental pressures. The concept of gods assume moral authorities however moral discerns common paternal intuits as external gods are illusory third party administrates. What governance assumes a supreme courts authority *"in opinions"* setting precedence to moral standards working opposite to natures principles in procreation is odious to social norm assuming a democratic societal acceptance. Homosexuality is a social/carnal perversion not a biological condition. *"Unless there are birth defects where gender is in question"* opinions to law are often based upon generational trend in expression to behavioral rebellion breaking any mold to status quo culture/traditions. Economic discordance plays major roles in social stage appearances. Better to be politically correct if you don't want to be criticized for offending anyone for their beliefs' people will simply join sides to defend others ideals and freedoms unknowing any illicit acceptance in social influence, if some wants to have same sex marriage what is the concern if there is compatibility though for most people it doesn't make sense, nor should it be a judicial involvement posing as ethical administers.

During puberty carnal obsessions set any precedence in thought to unnatural sexual orientation

Homosexuality is a paternal ambiguity habituated in carnal anonymity

Environment is the impression of cultural epitome perplexing comprehension postulating contribution conditions law in expectation; without evolution to technology there would neither be social conscience nor any concepts to knowledge of a collective is the cause to philosophy providing interpretation to common experiences as intuit compels definition. Perspective is essential "to individuality" though you have no control over what anyone assumes to know simply existing where there is no purpose as purpose assumes design and there is no design only life as we know it. Human perception; language has all origins in cultural relativity and literacy providing provisional linguistics to societal orientations.

There is no intelligent design there is no pre knowledge to conscious experience all culture has origins since cultivation in agriculture birthing material domestication in the establishment to community sharing tasks in any origins to crafts kindling economies beginnings.

There is no collective movie following any script the world view is any idiosyncratic feature

Belief in Gods is a unifying ideology wanting to make sense of existence assuming an authority of design

Materialism has sustaining properties

Confirmations are positive affirmations you are not alone knowing the world is often senseless

There is no collective conscious mind, media is a kaleidoscopic fraternity

Breakfast turns on the digestive tract activates cravings

Cultural interpretation distinguishes intelligence

Law puppets are trouble makers who do not know how to get along

Any world view knows an economic challenge in cultural disparities

It is best to not make projections having any expectation for successes and then you will never be met with any disappointment; dream worlds get the best of any minds anticipation

As children we assume parents know what they are doing to shield us and then we find later they did not remember their own childhood anxieties and inhibitions assuming in time a change to culture

There is no such thing as a collective mind there is only what you know the same in phases to adaptation conditioned to transcendence

When you realize there is no meaning to anything you will be alive, and then there is what you know as the force in transfer of instincts in survival subjected to a democratic delusion

What do you know? Describe it to anyone... A collective world view is an apparition

Central government authorities are ardent directors' conditions collective focus

Verbal pornography in television content has a degenerate censor

A collective aura is a strange postulation

EPISTLE TO THE CAPITALISTS BY TIMMY

The following is an exaggerated cultural profile to social subjectivity and expresses contradictions to Paul's ideology that god is knowledge

Paul Romans Chap I, *28 And even as they did not like | | to retain God in their knowledge, god gave them over to | | a reprobate mind, to do things which are not convenient: 29 Being filled with all unrighteousness, fornication, wickedness, covetousness, maliciousness; full of envy, murder, debate, deceit, malignity; whisperers 30 backbiters, haters of God, despiteful, proud, boasters, inventors of evil things, disobedient to parents, 31 Without understanding, covenant breakers, | | without natural affection, implacable, unmerciful; 32 Who knowing the judgment of God, that they which commit such things are worthy of death, not only do the same, | | have pleasure in them that do them. Chap II is an interesting confutation citing paternal causes.*

There is far more spousal abuse in poverty "works both flanks." Poverty has a higher rate to addiction and crime which fills up "corrections" facilities with people who were once innocent children who largely are from broken homes where life was often harsh and confusing lacking structure. (*What defines any causes to social sciences?*) Corrections to authoritarian rule postulates "*forced obedience*" is the cure however being obedient assumes authority to a "moral intuit" which has no definition or instruction manual "*only submission*" other than religious convictions postulating there are any deities who can do anything for you or anyone. No one knows your own story we get only glimpses of other's lives which are often different depending on any influences which subject us to change in personalities depending upon occasion. Assuming anything written in the bible is literal is atypical interpretation in those simply of faith or of contagious persuasion as power of suggestion plays upon the naivety in seductions to postulated virtues. If there was something to know we would all know the same interpretation being we are all subject to the same elementary experience in material conformity within an authoritarian barrier. Authoritarian aura has a temporal seduction in self-repression knowing its force to submission; authority is an ambiguous agent postulating ethics in social virtue.

Parenting where discipline converges to defiance challenging virtue initiates controls to codependence often transferred to God and law in frustration as a "paternal" authority

According to the Sermon on the Mount Capitalism in economic disproportion is malevolent

Any "pursuit" in happiness is a superfluous endeavor

The bible is so non contextual in so many fictitious accounts (allegoric expression)it's hard to imagine anyone believing anything being of true historic notability, nor do movies ever tell the full written tale/story of the Jesus character which in the book of acts states he was with only 100 followers *"after coming back from being dead in a tomb three days"* where then after giving some instructions to his apostles he was raised up into the heavens and vanished; pretty good magic tells us this is nothing more than a poets fable. And then why it was after his departure (why did he leave) did his following grow by the thousands; art is progressive always before its time is its metaphoric displacement. There is no intelligent design there is no precondition to consciousness only instinct in perceptive awareness to environmental adaptation, *"Human culture is not instinctual it is embryotic"* cultural epitome distinguishes any contextual knowledge, and then God is therefore knowledge, providing confirmations. The premise in the message of Jesus is one mind like watching the movie Pleasantville (*obviously an artist fantasy*) however, there is no cause there is no one mind there was no intelligent design and any other religious positions are relative to any possibility to an afterlife, which of course if such exists it would somehow be a condition relative to energies in sub-particle space or some alternate universe where any conscious life would be the same physical existence. The question is how would anyone know such a place exists tries any imagination, and if such did what would life involve? If such a thing as an afterlife exists one would hope it would involve a continuation to technologies innovations and perhaps parallels to existing culture though it surely would not be organic surely vexes the imagination.

The story in the book of acts is so farfetched it could only resemble a science fiction novel "a required read" suggests any movies ever produced on the bible have been plagiaries'

Acts I: 9 And when he had spoken these things while they beheld, and he was taken up and a cloud received him out of their sight 10 And, while they looked stedfastly toward heaven, as he went up, behold, two men stood by them in white apparel 11 Which also said, Ye men of Galilee, why stand ye gazing up into heaven? This same Jesus which is taken up from you into heaven shall so come in like manner as ye have seen him go into heaven.

Acts 2:44-46 And all that believed were together and had all things in common, and sold all their possessions and goods, and parted them to all men, as every man had need and they continued daily with one accord in the temple and breaking bread \ from house to house, did eat their meat with gladness and singleness of heart

One mind hippy movement doesn't work, socialism does

A true world view is what you know in the strife of what you and others go through to get by

Collective social focus in vogue successes is an enigmatic parade

You are your own universe is the sanctuary of your own devotion

Paternal senses get you right in the groin

Assuming purpose postulates intelligent design in a deity authority' there is no collective conscious mind there is no all-knowing eye in the sky

There is no collective mind in knowledge knowing our culture

Define liberty and freedom to privacy

Economy in development disperses innovative natures to technology expanding positions in forage propentence in transfer, inalienable rights are non-contextual, and a constitution distinguishes any physics to our cultural embodiment being equal in material cultures origins to our interdependence. What defines any parameters to privacy when capitalism has no boundaries in physical medium extensions expanding economic positions when there is no true purpose? What would law makers do for a living if they did not invent more laws to stay in business civil and economic? Law is necessary to provide any assurance there are certain boundaries however what provides definition to individual rights where constitutions provide no anthropological interpretation. Civil has gone a bit too far in the case of felony classification. Is a pipe bomb really a weapon of mass destruction or just a bias classification to an extremist culture, perhaps lawyers can earn a living changing a few laws so life isn't so harsh for certain people who can't get jobs due to their records offered only rehabilitation in prisons as correction to authoritarian obedience having no intuitive juncture. What is contextual to restitution, restitution may be addressing the causes to any caricatures offering rehabilitation and perhaps more authoritarian officials and religious ministers may become philosophy majors. If you are in a car with someone who was pulled over does the officer have the right to ask you for Id suggests a police state mentality or authoritarian despotism being owned by a postulated authority. Once you have any record you become classified stereotyped and profiled as the laws next contestant. Justice assumes disciplinary is restitution has no true nature is an authoritarian corruption. This is to simply give an example to how so many innocent people end up in jail with criminal records. Always ask for a jury trial however due to the increased dockets people are arrested with trumped up charges offered plea bargains lessens municipal expenses questions what ethical standards truly are in oaths to emblematic constitutions. Law enforcement program series "Cops" set precedence infringing civil rights due to social acceptances institutions despotism, where democracies cause preserving liberty is usurped having any potential class action suits against state and municipal jurisdictions.

"Plagiary" If you are innocent and there are no witnesses to testify or evidence of a crime no jury will convict you nor will a trial be scheduled and any charges will be dropped; this is especially true in any civil violations such as disorderly conduct trespassing assault or any other "mis-demeanor" charges. Social subjugation includes the pretentions to economic positions though nothing works if nothing works together being interdependent assumes nationalistic connections is very emblematic assuming any affiliation to any contribution

Any knowledge expanding education originates from philosophic interpretation looking through a rear view mirror in any attempt to understand and link any parts to the whole of cultural interdependence inquiries educations textbook revisions in certain categories where there is any science to knowledge, change in methodology, additions to discoveries and or changes to histories perceptions within what determines any authorities.

Social despotism includes class division which we are subjected to in transcendence by and large due to the barriers to economic apportion

Why do we buy in to anyone's opinion of us *"unless anything is true"* and what is true if no one knows our environmental influences to any dispositions

No one knows who we know or how we are with people, how do we truly know how anyone knows anyone

You are your own thing and what are you to anyone but any connection looking for anything in common is seductive and often allusive

You are insane to think there is any point to anything there is no point there is only what you do to survive drives people insane giving up a part of them selling out to fitting in is often a characteristic anchor to personality

The Band' Linkin Park; "*Giving up a part of me*"

Being fatigued can put one on edge is not the best time for interaction

Being out of physical condition causes lethargy lends to depression

Financial concerns affect demeanor in constant distress

Sitting for over extended periods restricts circulation conditions metabolism

I do not need you to be me do I

Dualism assumes any one knows who you know is image trajectory

Revolution is any evolution to human intelligence

Mark IV 13 and he said unto them Know ye not this parable? And how then will ye know all parables? 14 The sower soweth the word "Cultural context"

Context is essential to any knowledge including AI medium functions

Only the author of the Old Testament could have written the New Testament in reference to its stories proves it was written by a playwright/poet

Philosophic Hermeneutics is any practice to intuitive education and the most complex and yet fulfilling science to fruition comprehending experiences while writing anything accurately others would know. Movies parody social perversion so what is it and what are any transcendent causes? *"Insecurities condition inhibitions postulating a common virtue"* There are no intuitive minds knowing cause there are no authorities only casualness. Since the Old Testament begins as a fable *Adam Eve Cain Abel*, there is no true genealogy there is no biblical history, people are impressioned by so many misinformed documentaries they assume there are any theologian authorities. There is no problem, you/we simply exist and there is no cause there is nothing to know. Now do people think we are unprincipled when someone may be offended when questioning what they simply believe? If you know something you will not be offended knowing how you know what you know, no one can prove that Jesus Mohammad Abraham or any God exists, even the bible states no one has ever seen god at any time... Who's to argue if you don't have incentives for social principles you are going to have corruptions, which is the main theme to the Jesus story and one mind, as is the movie Pleasantville and in the book of acts *"They shared all things in common giving up all their possessions"* as if one could change free trade capitalism. God is going to damn you to hell doesn't make much sense does it if you are not good little boys and girls and Santa isn't going to leave you any presents in your stocking sounds like bribery.

John I: 18 No man has seen god at any time

The Old Testaments genealogy truly questions Jewish history, so was Jesus really a Jew in the faith of God Abraham heaven Moses and the Ten Commandments questioning his religion when after he was "suffered" ? by John the Baptist of what faith he then left to the wilderness to become a philosopher/poet.

A good question is' when did people start keeping birth records as in early times history had little record as any accounts were handed down as tales or folk lore.

Mathew 3:13-17 Then cometh Jesus from Galilee to Jordan unto John, to be baptized of him, But John forbade him, saying, I have need to be baptized of thee, and thou comest thou to me ? And Jesus answering said unto him, suffer it to be so now; for thus it becometh us to fulfill all righteousness, then he suffered him. And Jesus when he was baptized, went up straightway out of the water; and lo, the heavens were opened unto him, and he saw the spirit of god descending like a dove, and lighting upon him; and lo a voice from heaven, saying This is my beloved son, in whom I am well pleased 4:1-4 Then Jesus was led up of the spirit into the wilderness, to be tempted by the devil, and when he had fasted forty days and forty nights he was afterward and hungered and when the tempter came to him he said, If thou be the son of God, command that these stones be made bread (he then went to White Castle with Harold & Kumar) But he answered, and said, It is written, Man shall not live by bread alone, but by every word that proceedeth out of the mouth of God.

Believing a person died saving anyone from transgression is ludicrous, transgression is simply a fact of life and culture.

The story goes that Jesus all of a sudden showed up even though before birth he was already known as a messiah or King; (missing childhood history) and later in life he then went to John to be baptized; why? He then left into the wilderness for forty days and forty nights and battled a devil (*reprobate*; *was his mind corrupted by the culture*) was this his hermeneutic experience as an artist/philosopher? He then gives his **Sermon on the Mount**; this represents what any philosopher or any artist goes through to develop any talent or intrinsic art. And so why would he later be crucified if he was known to be the son of god? Yo brothers and sisters what's up with all that friction out there and these coworkers and students running around with semi-automatic weapons. What went wrong in all these people's lives? Please tell us again some guy died and saved us from social/economic transgression? Transgression has a constant and it is true that things in the past were worse than today now there are simply more people and so many more media's impressioning minds there is some epidemic while most are decent people wanting to live their lives respectfully responsibly in privacy and dignity. The world today is far more invasive than any time of the past due to the currents and trends to electronic Mediums of instantaneous information overload can make any ones world feel as though one is running on a tread mill being consumed.

Control Z is a macro on the word processer which retrieves what you just accidently deleted

Any addiction suggests anxieties

A simple book with detailed instructions on how to use and maintain a computer would be a best seller

There can be no intimacy with a cold fish living in a sea of antipathy

Organization & management skills are essential in anything one is involved

You feel stupid when you can't remember trivial periodic memorabilia

Economic recompense conditions human dignity

There should be a uniform code for recharging devices

Postulating democratic cause poverty is the seed to futility

Free trade capitalism is a maladroit incentive in irretrievable transfers expanding economic positions

Economy has an embryotic nature increasing positions questions conforming doctrines like "Credit Score" used in employable verification'

What do you know in happenstance what to talk about without looking stupid to yourself

Extended adolescence "20's" is a hostile environment in posterious endeavor wanting to make a mark

Natural inalienable rights are non-transferable freedoms, liberties and privacies

Is a bible taught preached or lectured in authoritarian diffidence

Complacence is the drudgery of repetition and familiarity wanting any creative spontaneity

The social pull consumes thought wanting to be a part of the world narrows focus. If you want to be good at anything it requires much practice, it has been almost nine years that the author has been in the study of cultural anthropology which one cannot do in forty days and requires one living in seclusion unemployed with limited relations. When you live in poverty you do not have all the fancy gadgets, ever watch the intrinsic art Get Smart... The thirties and the sixties were the best arts' today you have few intrinsic artists producing movies. Television programming has become so perverting with all the shows impressioning criminal minds is not intuitive art. Why can't you find; You Can't Take it with You, Duck Soup, My Man Godfrey, A long Day's Journey Into Night, Treasure on the Sierra Madre, The Fisher King, Pleasantville or Happy Feet, in main stream programming.

It's suggestive that you have no clue what you are a part of' art in metaphysics is progressive as dependence conditions advancement to an intellectual summons in providing any better understanding to what we are and what we do as one human culture. So what is the human condition? The main cause to democracy is liberties preservation and also concerns for the environment so what does determine liberal verses conservative in singularity to common ambitions? Are we running out of things to discover or invent as new and improved constitutes progress of innovations in efficiency and convenience. The origins to economic positions are relative to physical medium transfers in technologies providing any efficiency and or convenience. There are patterns to depressions which have not been educated that the sixties were in depression which caused much civil unrest. The depression of the thirties was far worse being there was no social security program then. Retirement at sixty would open more positions and it may be suggested that all corporations be required to offer *transferable* pensions being that minimum wages do not afford saving plans for retirement. What defines corporate responsibility? Go and have some conversations with some of those living on the streets many are not as injudicious as one may presume. What do you do if you do not have anyone to fall back on and you have nowhere to go? One can suppose we can simply let some suffer and let others die on drugs; then there is laziness wanting to be spoon fed in the disseminations to welfare programs which are easier for single women with children to be provided.

If you have ever had an obsession to use drugs you would know how horrifying it is to be an addict wanting anyone to come to your rescue, though no one can truly do anything to change the life you went through one can only shed new light upon any influence. *"Inadequate education for parenting"* where there is no instruction manual to our culture there is no constitutional interpretation knowing cultural probability. Today's physical scientists do not share any contextual supposition for defining Quantum Theory or Quantum Mechanics. How many movies are there in the past and today depicting Armageddon which seems acceptable yet try and imagine what youths today are going through as compared to what anyone experienced during the sixties-nineties being aware of the perpetuity to technologies impacts changing the way we think and the way we perceive the world?

What decides reason as was the theme to the movie *A Beautiful Mind* depicts John Nash as a delusional schizophrenic and that is often how anyone who thinks outside the box may be perceived; conformity assumes a collective world view; most people do not know how to have any dialog on the nature to our existence is competitive assuming competence is the superficial world of cultural memorabilia in trivial pursuits of who's who or who did what is nothing more than memorialized nostalgia. This links to how much education does not exercise comprehension makes people think they are not intelligent not knowing how to remember trivial knowledge. There is nothing to know it is how you know anything is definitive realization stimulates esteem knowing how to use our mind, artificial intelligence is the means functions intelligence is what you do with what you know being applied to your own successes socially and economically to one's own sustainability's subjective natures to capitalism.

Reality is a cultural idiosyncratic feature

Now days the media impressions a social esteem that does not exist so what is reality but a paternal malady assuming any goal in reality is any attachment assuming any identity in unison; there is no collective mind only any impressions to our culture is contravention however there is no interpretation to our culture following any plan because there is no purpose rouses greed in succession.

In the book of Acts they "knew all things in common" (*assumes one mind*) and gave away all their possessions infers Christianity would be based upon communal life or true commune-ism, *which does not exist anywhere*, and then there is the communism which is authoritarian police state/ totalitarian dictatorship in the compliance to obedience is a corporal perversion being owned as a slave by a thing in material bondage. In the bible is written the "*yoke and bondage*" which is the cultures tow chasing a dream for a perfection that does not exist it is all economic securement to our forage propentence transfer conditioned by the irreversible nature to free trade capitalism. Having something to do in life that makes sense is essential if you want to feel successful is a paradox postulating a collective ambition for common prosperity.

When you realize there is no point to anything you become one mind having no expectation and you simply exist knowing that you can never experience death being that if you were dead you would not know anything at all now what do you know? What does anyone know who becomes a politician when there is no interpretation to what we are?

Little Timmy as a little boy his step father for humor made him go cut a switch off a tree, he would and when he brought one back he told the little boy that it was not big enough now go cut another. He was resentful of him being the son of another man with his wife, procreation in domestication is a possessive obsession which does not exist in other species, "virginity *is sacred*" as procreation wants a certain purity in a mate for *duplication*; attraction knows likeness, is a cultural nuance, there is too much sex being promoted in movies which is not realistic concerning happenstance and duration, sex for most people is emotional sensual stimulation can be addictive. People often get married to have their own concubine as the world is very intimidating wanting to be a part of someone or something to make one's life complete. Marriage bound by law changes everything; compatibility is friendship sharing common interests has significance to lasting relationships.

Connecting our minds is the main theme to the bible "assuming" such a thing as one mind. Authority assumes directive. There is no one responsible for anyone so there is no obligation we are all dual diagnosed we are addicted to our material requirements subjected to the news perversions and we all may presume there is a collective mind keeping track of anything so what is real and what is there to know about anything. One mind assumes there is such a thing as a collective mind following any plan though what impressions our minds is technology evolving providing futuristic views in the ardent tides to progress.

Nationalism is the idolization and memorialization to idiosyncratic memorabilia assuming a national character in the perseverance to an illusory cause knowing nothing as it was being no one was ever there knowing how to describe the influences subjecting any minds suggestions to cultural motives being there is no true history knowing casualness is cultural anthropology or the matrix existing without an interpretation is the cause to philosophy.

Mind to moral ethical standards weighs consequence in reprisal

Despotism is the loss to liberty influenced by economic depressions

Zoos are a cruel amusement as human keepers of the planet

Children with no traditional family having two biological parents has a sense in abandonment

The ignorance of war causes much suspicion where anyone is suspect

Dysfunction in family (paternal malady) has an embarrassing stigma in social comparison

Pretended knowledge in religious apparitions has a posterious envy in kinship

Everybody's right everybody's wrong when nobody knows what's goin on

Existential polarity assuming a social view' no one knows who you know

Do they reason in Chinese or do they reason in culture

Cain's malevolent persona has its own insecure disposition

Notes:

Section Ten

Children are a permanent extremity carrying much responsibility being attentive can be very consuming fostering any patience, children can sense any distance, having children when you are still a youth yourself can be very frustrating when you have not had the time to experience life on your own maturing clarity to one's own social anticipations and desires for successes, kids are a full time occupation and stimulate many emotions which are centered upon the nurturing to the parent child relations where there is little patience lends to the distortions of authoritarian ~ chastise which sets precedence in patterns to invectives. In the art of parenting having respect for common inhibitions to social infirmities playing the suitors advocate is a practiced intuitive art. Where does anyone attain esteem if not by the principles to their conduct if not feeling any sense in appreciation? A child wants a parent to show an interest by being involved which is often limited when the pressures of life's many responsibilities lends to being self-absorbed. Organizational skills having structure is the most essential practice for preparing children for life's terms in self-manageability to common sense practical applications. Managing time with daily routines while having behavior charts with earned rewards isn't a bad thing when dealing with unruly kids ignoring bad while praising courteous behaviors sequesters intimidating conflicts. Bribery sets any precedence being controlled in pacification.

There is much pressure in youth postulating consensual age is 18 creates much pressure which is consuming when you realize no one can do this for you and it is likely worse for females who are expected to be care givers enters perhaps the cause to women's suffrage wanting to have any sense to independence. The best escape from the parental clutches for many is finding a partner and getting married to have someone support them. Ladies by domesticating traditions are also less mechanical and are most accustomed to be home makers/care givers for children while wanting to be taken care of by a husband, although today in times requiring two incomes there is often a double standard in these roles while at present there are many positions for women in many fields of expertise due to woman suffrage. Girls just want to have fun and so do boys with the boys in their own fraternities.

Raising children has intuitive reflections reversing roles' authoritarian has a certain lack in attentive patience and then there is any balance between the two

Children having media devices at early ages may not be the best thing contributing to distractions' don't give in to peer pressure' a social addiction "**media devises condition attention**" limits concentration at school

Video games causing over stimulation contributes to aggressive children

We are conscious observers most to what we go though is non contextual and superficial. And there are the layers to capitalism providing positions. The world is inhibiting due to all the expectations which are placed upon anyone wanting their children to be successful. Successful children assumes being successful as a parent and then there is what stimulates motivation being your own thing in discovery. No one can do anything for anyone; paternalism— children sense parental anticipation assumes the interior reflections to identity. Kids appreciate you when you are not condescending being real. It is less likely that autism is a genetic physical disorder and it is true that assuming so causes a stigma centered on the idea that someone's brain is defective can cause a corporal stigma. The same is true for ADHD. Consciousness is mysterious wanting to know how to apply your mind to anything constructive. Consciousness knows cultural impressions. Autism would be linked to how conscious observation construes a relationship to existence developing comprehension to language though the world is pragmatic postulating a collective known ambition. Autism is likely relative to an acute manner of perfectionism using construal knowing what anything is or does has a greater paternal sense to social antipathy.

ADHD is environmental conditioning not a physical defect, diagnosis in the power to suggestion can and does cause a stigma in thought as being physically defective much as suggesting that alcoholism is an inherited genetic disease. Children do not know a parents world view nor does the parent know what the child senses thinks or assumes. Much thought concerns how we perceive ourselves attaching to any identity within any associations to other personalities. Is it possible to comprehend all causes to autism, our culture is very dumbfounding and we are born with intuitive curiosity in a world of self-embellishment; intelligence is wanting to know what you are being constructive applying the mind to any art to discovery is very competitive; memorizing things as nouns is not intuit. Autism likely would be caused by an acute emotional sense in fear of cultural adaptation in social relativity and motor function in a world conditioned to material accessory. Codependence caters to emotion empowering dual insecurities to paternal senses of inhibitions to intuitive dialog or affirmations' confirming common idiosyncratic natures is responsive. One can be consumed by rhetorical thought prompted by the things we experience and do not know how to respond to intuitively interacting motives to resolves and retorts.

Now what is authority in any assumption anyone knows how to describe what is going on and what is it anyone knows how to describe other than a competitive nature wanting to enjoy life, assuming there is some collective eye knows who anyone knows is existential dualism, no one knows anything there are no instructions to the game positioning any outcome. Reality is the physical realization to existence (the human being) and the social implication assuming anyone knows what we are doing involves cultural anthropology. Losing your composure to aggressive thoughts and behaviors is very demeaning exposes any vulnerabilities to social compositions; boys are very competitive contesting courage is a corruption provoking retaliatory responses, movies impression there are many social infirmities.

"I fight authority and authority always wins"

~ John Cougar Mellencamp

Doesn't everyone want to be a rebellious rock star?

You are just expected to get a job and then there are numerous discriminations, age, appearance, experience

There should be more emphasis placed on how children are addressed to stimulate intellectual conversations rather than baby talk. Paternal pandering is emotional subjectivity and it can be so difficult to describe emotions or the causes to any expectations for perfectionistic senses, cultural linguisticality is largely superficial however, inhibitions are true for anyone wanting to feel successful. Adults are still like children competing for approval. Mothers assuming to be the paternal expert (by tradition) postulating intuit insinuates superior domain in the spousal relation rouses animosity. It is best if both parents are equal is sensed by children supervenes rivalries caused by any division ensues competition. Wanting to have a productive role is competitive wanting to be approved socially. "You wait till your father gets home" is a transfer in frustration having any uncooperative child testing anyone's patience is an authoritarian transfer.

The arts presume we are all part of the same game, though we are not when we have no concurring directive conjures religious beliefs

There is no interpretive cause limits constructive thought in appreciation to tailored skill

The culture is superficial where there is little to no contextual interpretation to much subjectivity, comprehension reflects and admonishes any experience is literacy; children sense people's emotions and insecurities especially an over bearing parent, perfectionism infers defect, speaking to a child should be the same as an adult non condescending; baby talk gibberish associates incompetence much like fairy tales. Wanting to be approved is an underlying inhibition though what is considered success' "adaptation to a material expectance in self-sufficiency imposes a paternal obligation" Cognitive for mainstream presupposes organizational and personal management skills which also include individual relations as well as what we are a part to as one global culture assuming a democratic obligation. What is the primacy to democracy is it social or economic one conditions the other?

Confidence is gained knowing how to respond to subjective experiences with corresponding insight; Intuitive cognition knows commonality, body language and verbal tone have significance in knowing how to communicate any anticipatory involvement. Catering to insecurity fosters codependence. Inhibitions sensitivity is transgression posturing domination is extremely competitive. Another layer to the Cain Abel metaphor Cain slew Abel due to being jealous being insecure wanting to be approved assuming a social mind knew him for his proficiency or success; he sensed failure when his illusory god did not approve his countenance, a social view in dualism. Dominance assuming competence is a maladroit conclusion and a capitalistic perversion. The sense of shame is any sense in accepting any inadequacy. Physical coordination is our first challenge which is also met with visual accessories as stimuli having no contextual function simply amusing gratification. This would be the primacy to any addictive tendencies suggesting reward. Over embellishment with children is foolish and a means in purchasing affection by unearned rewards, cancel Christmas. The economy would collapse. Do we reward children for simply getting through another year? Cooperation charts work is good practice for encouraging responsibility prepares children for being self-reliant. Organizational skills prepare children for school knowing how to manage thought and time. Children should have basic reading and math skills before kindergarten so they do not sense inadequacy and fall behind the teacher's pet that is often praised and paraded senses exclusion. Definitions for many words require practical applications such as the physics to mathematical implications which adults take for granted. Education is socialisms gate nationalism is an idiosyncratic suggestive trait.

~The Grinch

When there is no contextual application we do not use intelligence, intelligence knows what you are doing; when you are distant to a child means something' you have no clue how to communicate your concerns to what you yourself have experienced or are experiencing as inhibition transfers Adams transgression to Cain, is the same metaphor as Memphis to Mumble in the film Happy Feet tells us there is no One, there are only artists and philosophers whom are metaphysical scientists. Attentiveness with children conditions a bond in assurance we have common senses. The precept to consciousness is that there is nothing to know being that there is no intelligent design there is no intent so there is nothing to measure. It is not what you know it is how one knows anything using logical assessments, however what anyone knows is how perverted the force to what we do lessens the experience to any art in practice, there is no collective mind recording any knowledge unless there is anything written down in books relative to what anyone has known. IE the premise to bibles' however metaphors are often incomprehensible to most people.

When children have Down Syndrome people may speak to them as if they are retarded which is very perverting where our culture is all about visual perfection is a great example within how condescendence can be oppressive. There is no intent having any collective goal reasons that there is nothing contextual we all know within how anything applies using intuits which is what stimulates esteem knowing how to know anything. Intelligence wants to make sense that we know what we do is constructive and then there is the Cain metaphor wanting any approval is competitive in a contest assuming any goal. You have always been the same person saying what is this this is stupid enters law. Law is a reactionary discipline control device assuming perfection to a moral conviction. Fear can and does condition neuro transponders relative to constant cogence. People are essentially superficial because they often have no clue how to translate emotions is often inhibiting sensing any insecurities. Any speculation to a collective conscious view is memorialization to cultural idiosyncrasies.

Consciousness is the voice in your head asking questions many to which we do not have all the answers

The revolting child is one who does not understand why their parents do not know what they are experiencing emotionally in adaptation to a world that is often malicious and difficult to get along in is the origination to what often has been referred to as bipolar where there is much paternal animosity "distorting" responsibility. There is much inhibition as early as three (a fog) wanting to apply our minds to our experience in comprehension being that we have little language skill we would sense isolation and have not yet experienced the illusion to a collective mind which has any origins in the nature of codependence assuming authorities. Education impressions there is an established collective mind to know is selective and bias to conforming idiosyncratic institutions. Socialism is the transfer in responsibility to a collective paternal mechanism. Comprehension applies any individual application to any practice. With the grading system education is very competitive assuming oneself incompetent even though what is being educated is still largely pragmatic knowledge; Intelligence knows how to comprehend anything experiencing comprehension is the practice to intuit reasoning by deduction any contextual application. Emphasis on language use should be the first mastery in order to comprehend any other instruction. What are any functions to mathematical division and multiplication? How can anyone know mathematics if there is no application or comprehension to its language? Nouns are not true physical comprehensions though still require memorizations to applicable definitions in practice, practicing anything consistently conditions auto recognition. The medium is the artificial intelligence to our cultures functions as the existential experience in conscious subjectivity applying any comprehension. Without technological evolution we would be nothing more than primate's still living in caves applying no intellect in the subjectivity to our cultural experience assuming any interpretation is quantum physics.

Mathematics is quantity, volume, velocity, pressure and geometric measurement using formulas in pragmatic applications is an educational focus postulating memory in logical comprehension is intelligence

Intelligence is intuitive relativity to culture

Artificial intelligence is the means intuitive functions

Measuring success wanting approval has a left temporal polarity in ego social reflection.

Seeing how others are being controlled by a person is the most difficult dilemma knowing it is a process of submission on the part of the victim whom often gathers attention through the drama of martyr; that might sound a bit harsh however control does come down to self-inflicted reflection of esteem and worthiness, it's not often expressed well in movies. Have you ever been concerned for someone you know who has anger issues? There are still the variables of conditioning often forced by the fear of physical submission and of the significant others insecurities. These conditions are often enduring situations almost always tied to financial securities of material dependence and it is not gender biased. It does question the concepts of equality relative to the controls which gender often influences within the means of law and order. All social issues begin in the home which influences the impressions of children's perception in the course of contravention by and through example and it is a real fear which is experienced effecting environmental worthiness which gets buried deeply in the child leading to patterns of deservingness in either direction. It is not poverty biased other than the poverty of empathy in the understanding to the common principles of conduct which induce a sense of assuredness.

Love is an intuitive connection where there is no possession is loyalty and confidence, conquest seeks approval and then there is buyer's remorse some women like to seduce playing a game of dejection being insecure wanting dedicated assurances, a woman wants a confident man a man wants a secure woman.

Drug tests; having a hangover is more dangerous at work than testing positive for THC where there is no residual affect altering coordination or cognizance, in fact THC stimulates awareness. Drug testing is discrimination and a bias opinion alcohol is a poison and is the cause to many social perversions/distortions. Authoritarian and paternal order assumes a collective preeminence though subjectivity is a capitalistic conditioning to environmental opportunities sequestering imagination. There is no science to law there is only science to causal determinations disseminating liabilities postulating restitution knowing cause.

Money in expenditure conditions time

There is less conspiracy than you know plays any part assuming authority there is no connection where there is no goal assuming a reality anyone knows, capitalism subjects us to any appreciation to any contribution assuming any obligation. Nothing works without capitalism so what is reality but incentive.

A disheveled house is a pack rats haven keeps a mind in motion

Social embarrassment has a cloaking stratagem to memory

If drinking and driving are against the law how are taverns legal

Indecision has no constructive calibration

Consistency conditions memory in cognitive manageability in regularity to functions

Ignorance defies the nature of reason

Thou shalt not defy the laws of intuitive nature

Money is very stressing for anyone especially elderly with fixed incomes conditioned by inflation

Foster homes have a sense of abandonment in estranged dispositions

Authoritarian has no intuitive virtue in memorialization

Overhead expenses are all inclusive to inflation

Elli May Swamp Mountain Spring Water has genuine marketing qualities: Patent Pending

Authoritarian subjugation has an estranging polarity in submission

Moon Shiners have been around for a while Now Premiering on Discovery Channel following Amish Mafia:

Give a Hoot don't pollute is too convenient in road side corporate franchise advertising

Notes:

Section Eleven

Diagnosing ADHD; the challenge in medicine is that doctors are often scripted "unknowing" the actual condition (s) which cause "hyperactivity" or "attention deficit" attention to what? Parental expectation shoulders a motivation in personal responsibility where there is any force there is no art to development, the problem with education is any postulation intelligence in comprehension is only relative to good memory; however good memory recall is relative to contextual experience and repetition; our culture challenges intuit wanting to know our own relationship in receipt or involvement questions what is there to know in the shades to postulated authorities where cause can only be distinguished as self-sustainability. The reason nouns (names) often have no intuit memory is relative to the fact there is no context in relativity other than idiosyncratic features; other nouns assign memorized functions" run, walk, rain, drive to sound association. Hyperactivity is caused by complacency, uncontrolled thought and an over whelming desire for stimulation where there is little to no amicable structure or art to craft where any practice rewards talent distinguishing personal value. Scouting programs do provide a social venue and any instruction rewarding achievements while being a part of community as does the participation to any sports which may reach any limitations due to time and financial abilities which can be overcome by community center programs receiving any investment from financial contributions and volunteer mentoring programs.

ADHD has everything to do with the frustrations to social abilities combined with constructive talents in the adaptation to material domestication wanting to be successful while expected to be in the anticipations of reward which stimulates ego in memorialization in the positive or the negative, ADD is relative to rhetorical or consuming thought where experience had or has no resolve, distractive thought can restrict comprehension while reading where distraction may be more relative to the definitions to formulas in the use of words forming a story or developing an understanding to any relationship IE as in the logical functions to mathematical equations. Consumed thought can also be relative to the pull of the means to any expectation in self-sustainability. The most fundamental aspect to education is language skill in the linguistics to any course where the comprehension to relation develops any scholarship, and then there are the social distractions of recess wanting to be on a team, children want to compete and win the game., If a child has decent reading skills before kindergarten and first grade they will have more confidence in the ability to learn and not fall behind in the syllabus and after all the means functions are AI in any specialization. Another aspect to education *"may seem redundant"* concerning comprehension is relative to the fact that memorizing names to things in distinctive identities exercises no intuitive construct or actual practice to comprehensive competences; any knowledge concerns the interpretation to relations. Agricultural despotism also includes the labeling of plants geology geography and other species in categorizations conditioning cultural perceptions..

The question is what does a parent require to know to raise a child and then of course there is the question of gender and roles to domestication in the assumptions to care giver and provider where the two seemed to have merged in the requirement of two incomes to afford any life style. The truth is that competition to cognitive skill is most relative to efficiency stimulates leadership qualities can be quite challenging contests competency in mechanical skill specifically where there is a communal goal for achievement. Having any chosen talent at any age stimulates any esteem in confidence stimulating any precedence in failure seeking any perfection is art where there is any definition. Children want to receive any approval in ability always seeking attention from parents and peers stimulates competitive natures. What is going on around us in family wants any insightful consultation having any sense to inclusion is any prevalence to social skills.

The number one reason people use drugs is to alter a consumed state of mind, and what are the causes to obsessive thought, *feeling like a failure*. It should be obvious that everyone suffers from some degree of consumption with all the pressures; lets face it we live in a very confusing world subjected to much symbolism postulating any unity. The key to renewal is in seeking knowledge to how we are conditioned by our culture in anything challenging comprehension affects the way we perceive individual and collective intelligence subject to the assumption of authority conditioned by law. The fact is that there is no human cause only casualness there is nothing to know it is how you know anything… *"All these little things in life they all create this haze" Running out of Days'* Third Eye Blind (The Band) The enigma is that people are inhibited not knowing how to communicate their true emotions and the fact is that we all have common experiences due to the adaptation to our culture which has had no definitive instruction. When we are children we are curious beings asking the question; what does anyone know challenges any sagacity where people are simply scripted to trivial memorability and social/cultural innuendo? Intuit is the comprehension to relativity social and economic assuming any collective directive is the cause to philosophy in providing any interpretation to "Cultural Anthropology" or rather what conditions the existential human experience is psychology in the adaptation to the domestication of our culture. When we are children our view of the world has less pressure being that we are living in the nest of our parents material procurement that view is mystical unknowing what anyone knows and what culture does.

Transgression is well depicted in Pay it Forward and The Fisher King

Being deceptive' tone in seduction is convincing "Our Father" postulates a paternal guardian plagiarizes philosophy

The bible is an existential allegory to human culture

Reality is a convoluted question assuming there is some collective conscious mind recording everyone knows and follows knowing all the positions assuming any collective motion picture to time; reality would assume knowing any relationships however there is no interpretation to cause so the only true reality is what conditions any consistency in the physics to our culture in material adaptation and conformity. The truth is that there is no definition to any cause there is only any casualness to which conditions human behavior. What conditions reality is human innovation in a provisional sense of new and improved mediums of modern convenience conditions any projection to futuristic landscapes. When we are young the question arises what is there to know in the vast nature to the physics of technologies like reading popular mechanics wanting to know how electronics function challenges any comprehension in the sense that instinctive natures in survival due to materialism are conditioned to AI which require a certain pathology in knowing any ancestral origins to any evolution knowing any fundamental knowledge to physical relationships advancing.

Social and economic transgression conditions all subconscious culture in dyslectic archive.

There are no two perspectives sharing the same world view

Innovative efficiency convenience is a prepackaged salad with dressing obsolesces craft

There is no independence there is only economic interdependence

Codependence has a covetous polarity; Linkin Park "Giving up a part of Me" I let myself become you…

By and large education programs memory

Mind has no common directive other than freedom bound to an interdependent culture converts to financial liberation

Opinions of others has a strange consultation with personality weighting candor

Frustration and Anger can be likened to a toddler's disposition

The movie 10,000 BC 2008' in domestic cultivation religion has exploitive connotations

Commitment has a very strange sense in polarity

Authoritarianism takes hold when we are young looking for any shield of defense *"God or law"* in retort to facetious behaviors of dominating people who want to control others with fear as a test of courage; IE transgression stimulated by the competitive nature to capitalism and cultural domestication, it's difficult to respond intuitively to things that don't make sense ; one of the key factors to transgression would have anything to do with economic placement being a force rather than any art to discovery or innovation stimulating any creativity. Transgressive behaviors can be attributed to insecurities which would be an aspersed aptitude in economic/social retaliation or rebellion conditions any animistic polarity. The truth is that any social polarity can be and is very inhibiting where any pride will be conditioned to having any talent or specialization distinguishing any esteem in being successful or accomplished projects any posterity in reflection shining on the world in the imagination to a self-portrait to characterizations in comparison or measures to identity. The linguistics to language distinguish any relations to social and economic subjectivities in conformity to the cultures interdependence in the transfer of human innovations in technology conditions human existence in provisional expectation to contribution in pulling one's own weight affects any social antipathy or prejudice. Existentialism presents any subjectivity to comprehension in identity to any relativity. The truth is that intuit is any ability to deduce relations social and economic though economic is provisional intelligence (AI) connected to survival is environmentally impressioned and selective to opportunity and interest.

Any national union is nothing more than symbolic idiosyncrasies fighting an invisible war in ideological postulations conditioned by what technology does affecting culture expanding economic positions, technological evolution conditions the presence of time in changing fashions to human accessories.

Fists or mediation has an estranging fatherly virtue

High School has the worst division in social polarities

Reprobate mind is a social orientation to a competitive friction

---

"The Prisoner Series" 1968 Social rebellion is a sign of intelligence

When you're praying who's listening

When you are praying who is listening

"The Wizard of Oz" 1939 has a strange consultation with courage heart and mind

Indecision has a strange consultation with perfection

You are who you are no one knows, there's no place like home

The fact is that much experience passes through the mind in a fraction of the moment where there is no mediation providing language in self-dialog comprehending any relativity to association, the pull to our material bondage conditions any mind in projection to self-expectation wanting to be successful in any sustainability to any provisions to our culture where anything can be turned into a business or license playing upon any instincts of survival subject to the conditions of free trade capitalism. The truth is that the whole world is a conspiracy because there is no such thing as a common cause assuming any authority knows any directive. What drives science is making sense to physical existence challenges human intelligence which cannot be argued where there is any truth. The fact is that any assumption to reality is any connections to the posterity of any inheritance to traditions in expectations to any assimilation to cultural idiosyncratic environments.

Every economic position has an interdependent value

"The Fiddler on the Roof" "Tradition" 1971 represents a cycle in despotism inspired by the 1960's social rebellion'

"Duck Soup" 1933 is an expression in despotism reprisal

The ultimate question of science is where are we going "Quantum Theory" "String Theory" Does matter extend to another universe

Social dejection has any sense to humiliation

Being constructive keeps any mind in tune...

The mind in perception is a physical recording in peripheral existence and environmental experiences; real time reflection and projection which conditions analytical and rhetorical thought contemplating and defining relativity conditioned by the existential experience in cultural subjectivity and instinctive natures in self-sustainability. Recollection circumstances subconscious memory in experience in prompts and repetitive dispositions.

Don't dogs have brains, why can't they talk? The origins to how human language evolved: humans have certain physical attributes giving them certain agilities to invent conveniences in technologies which has all origins to an expanding language as nouns and any comprehensions to relationship which is what conditions conscious awareness and any want to communicate experience was once extremely primitive though evolved due to the material culture stimulating human interdependence as crafts stimulated barter eventually evolving an economic condition inventing capitalism.... enters AI and the means functions expanding the perpetuity to education providing instruction which is the cause in philosophy providing knowledge relative to any cultural domestication in evolving material interdependence summons control mechanisms stimulated by the cultures corruptions conditioning authoritarian subjectivity in the invention of law conditioning a sense to order...

The truth is that most people simply do not know how to read the metaphors in the OT and NT and if they did they would realize the bible was written by a poet allegorizing our culture. *Cain sought approval from his illusory god didn't receive any became Roth slew Abel and left to the land of Nod*, consumed thought and distractive thought is relative to how "much experience" has no resolve conditioning memorialization in a myriad of cultural subjectivity in the adaptation to a material domestication having no intuitive instructions subject to the maladroit obedience of authoritarianism conditioning any expectations to conformed views of indigenous heredity, and especially religious views in pretended knowledge which have no tangible conception. If anything can or does exist anything would be self-evident. The fact is that law imposes a moral conviction law is therefore a religion assuming a higher authority (invention of a creator knows any design) there is no separation between church and state, Supreme

Court justices establish laws based upon opinion construing any causes applying any intuitive knowledge to the cultural epitome where there is no intuitive education to the cultures conditionings in conformity to authoritarian rule where any law giving anyone any rights infers proprietorship. Law is conditioned by contractual omission regarding why signatures are required, you have the right to remain silent anything you say may be used against you in a court of law by omission postulating authority, however a jury has the right to decide any law, "paternal law" people of government are simply representatives, law is the law of the people.... Jurisprudence in the dictionary is defined as the science to law however there is no science to law, law is a reactionary device and an authoritarian clause.... a form of transgression or a non-intuitive retaliation where in actuality people are expected to be accustomed to a maladroit obedience in submission as a form of correction, Jurisprudence is the prudence of the Jury to weigh the penalty of offence based upon any evidence submitted and constitutionality of law. Criminal as defined by the Gilberts Law Summary is the intent to harm another person or their property. Using drugs for instance is not the intent to harm anyone however can now be considered a felony is opinion is simply another label in capitalism; it is what people do while on drugs initiates criminal acts or illicit behaviors. Lawyers are in the business of inventing laws pretty soon it may be illegal to question authority where there is no intuitive comprehension, "despotism" this is no way condoning the use of hard drugs; drugs get you nowhere the question is why people use them if not to escape an imperfect world of expectation and confusion. That stated law is still a necessary appliance without law we would have a more barbaric world, the reason we have law to maintain order is relative to a material interdependence. The fact is that there is no actual constitution providing interpretation to what has been evolving our culture is the cause to philosophy defining cultural anthropology to understand the nature of our existence and also why the freedom of speech exists.

Reality is culture culture conditions reality

Democracy liberty & justice strives for equality strained by inflation

Females in poverty have certain vulnerability in options lending to domestic abuse and addiction

"Without permission and by what account, police body cams violate rights to privacy" you have the right to remain silent there is no cause there is no law only contractual omission by signature

Deeds are the company you keep

Transgression transfers through crime stimulates authoritarian despotism

Consensual age having to leave home is a lot of pressure

Authoritarian despotism in an accumulation to law has any sense in loss to liberty has a culmination sparking civil unrest/rebellions IE "Ferguson MO and others" much like the sixties rebellions. Law is also stimulated by health safety concerns caused by increasing protocols to technology such as seat belt or cell phone use, restrictions or smoking ordinances as authoritarianism is an existential paternal paradigm as a cultural custodian watching over your shoulder. Race ethnicity and economic prejudices play roles in stereotype profiling is repression in the opposite of laws objective for preserving individual liberty and postulating a paternal fraternity of one mind to principalities.

Every jury in the land is tampered with and falsely instructed by the judge when it is told it must take (or accept) as the law that which has been given to them, or that they must bring a certain verdict, or that they can decide only on the facts of the case.

"If it is possible that such a practice as that which has taken place in the present instance should be allowed to pass without remedy, trial by jury itself, instead of being a security to persons who are accused, will be delusion, a mockery, and a snare,"

Lord Denman, C.J. O' Connell v R. 1884

Trial by Jury is a sacred trinity of words that means a person is judged by twelve peers. Their verdict represents the common sense judgment of ordinary people and is based on right and wrong according to each jurors own conscious. It is their duty to administer justice.

"Justice for all was a principle they understood and believed in: but by "all" they did not really mean persons lowdown and no good. They meant that any accused person should be given a fair, open hearing, so that a man might explain, if he could, the appearances that seemed

to be against him, If his reputation and presence were good, he was presumed to be innocent; if he were bad, he was presumed guilty. If the law presumed differently,

The law presumed alone."

James G. Cozzens, The just and the unjust {1942} 57

....It is presumed, that juries are the best judges of facts; on the other hand, presumed that the courts are the best judges of law. But still both objects are within your power of decision... you have a right to take it upon yourselves to judge of both, and to determine the law as well as the fact in controversy,'

State of Georgia vs. Brallsford, et al 3 Dall. 1 {1794}

If a juror accepts as the law that which the judge states then, that juror has accepted the exercise of absolute authority of a government employee and has surrendered a power and right that once was the citizens safeguard of liberty.

For the epitaph which can be carved in memory of a vanished liberty is that it was lost because its possessors failed to stretch forth a saving hand while yet there was time.'

Sutherland, J., Assoc. Press v N.L.R.B. {1937}

Jurors have it within their, as occupiers of the most important decision making office in the land, to nullify every rule or "law" that is not in accordance with the principles of nature, god given, common or constitutional law. It is this power of nullification that makes the **Trial By Jury** one of our most important rights. It is the one and only right that can protect and preserve all of the citizens other rights and liberties through peaceable means.

"I CONSIDER TRIAL BY JURY AS THE ONLY ANCHOR EVER YET IMAGINED BY MAN, BY WHICH A GOVERNMENT CAN BE HELD TO THE PRINCIPLES OF ITS CONSTITUTION."

Thomas Jefferson, Letter to Thomas Paine {1789}

*

You should have the right to a jury of your own peers in those who know your character

People do desperate things in desperate economic times due to the perversions of capitalism vying for survival

Transgressive behavior in rebellious adolescents' subservient's teachers in aggressions domination ready's arms with guns in school

Cultural adaptation to material domestication in adolescence has a transgressive nature in comparative dominion

Inflation causes greater divisions in human distinction mandates socialist applications by and for people

Liberty knows a classless society

National emblems cast idiosyncratic invocations

Job qualification is one of the most self-degrading comparatives

If there is an expectation to human principality/material quality environments societal epitomes

Seat belt laws are an infringement on personal liberty coerces probable cause is a police state mentality

Authoritarianism is a reprobate mind in cultural perversion

Trust has no urgency

Using a dictionary and thesaurus practices comprehension

The truth is there is no collective conversation knowing any true cause we are simply living within the bondage to a material culture where we must adapt to the artificial intelligence of our means functions which requires an advancing education. Socialism; do we expect education to raise our children; the fact is that children who have no reading skills before kindergarten fall behind and become classified, literacy has everything to do with functions and social skills in the comprehensions to any experiences. Praising children for memorization is non intuitive' memory alone is not comprehension to contextual relativity. People who memorize trivial pursuits like watching Jeopardy assume memory is intelligence to cultural memorabilia. Most education has no practiced or practical application where there is no art to exploration or discovery where the world is all laid out for us to fall into where it is not what one knows but who one knows given any opportunity requires social skills and the aptitude to ambition wanting to be successful. The brain has no intuitive function the mind is intangible subjected to interpretations of the existential impressions, the brain is a neurological system of motor functions conditioned by the minds reactions of the physical and metaphysical experience, the transgression of our forage propensity conditions the stressors which condition rhetorical thought where experience has no resolve contributes to attention deficit, the truth is that people simply become programmed to the day to day tasks of material maintenance and occupational robotics where the mind becomes subject to familiarity limiting exploration though in reality there is nothing to know we simply exist as conscious beings wanting to make sense of our existence.

Culture challenges intuit conditions speculation where there is no reflection or mediation to cause in retorts conditions any memorialization which has an accumulative consuming affect conditions any compulsive disorders and causes to addiction

The New Testament is an expression in sociology and despotism

King James-Shakespeare-Socrates-Plato-Epistles-Meno-One mind-Virtue

Culture conditions philosophy

Thinking is reasoning relativity

Having a regular 12 step meeting is a good fraternity; Corinthian I XIV

Leonardo Davinci b. 15 April 1452 d. 2 May 1519 was revered for his technological ingenuity he conceptualized flying machines an armored vehicles concentrated solar power an adding machine and the double hull also outlining a rudimentary theory of plate tectonics

The only way to know what the Roman Catholic religion was prior to Jesus is by historic Vatican Archives

The Roman Catholic religion adopted the gospels of Jesus Christ after 1611; Christianity had no prominence until King James

Movies and film documentaries "impression" any past or present collective perception of time anything biblical is fiction

Religion has any indigenous misconceptions' a Jew is not ethnicity it is a heredity in traditions; the fiddler on the Roof is an expression of despotism

There was no media before the printing press as any knowledge prior was a crow's crier

There is no intelligent design' culture is conditioned by human ingenuity

Cain and his plow metaphor origins to cultivation birthing settlements in material domestication

There is no collective state of mind' global conflict impressioned by media conditions any bias polarity

All violent offenses stem from a deep sense in dejection "the Cain metaphor vagabond" postulates exclusion to an illusory circle

There was no written history before the printing press (1430) has significance regarding why people had read the bible literally postulating creationism in the mystery to human origins

Notes:

Section Twelve

Is it impossible for god to exist? "Yes" there were no physical formations in the universe before the big bang occurred if there were there would be artifacts; all life evolved through any evolution to single cell organisms, right after the big bang all planets were globs of molten elements which changed form from liquid to solid matter, as the earth has been cooling water was vaporizing and separating, condensation conditioned weather the atmosphere is molecular. The earth was once more uniform or a smooth texture, as the planet cooled there was crust which was being cooled by the stratosphere, this formed plates which meshed together forcing them upwards into mountain regions, like cracks in the crust (look at the Sierra Nevada's); think anything through and add to the statement, (geological hermeneutics) now the question to the evolution of consciousness would be relative to any species subject to the adaptation in a changing environment? There is simply no way to know all the physiological occurrences to know how consciousness has been evolved though may be more related to the instinct of survival subjected to a changing environment and an increasing food chain. What does a porcupine know or chickens who cannot fly too well and always seem to be a bit insane in the way they strut assuming confidence. Chickens are pretentious as much as humans can be when they feel like failures.

Nothing comes from nothing

Consciousness is cultural awareness evolving

The question of creation is ludicrous as if there was some divine being who would have known how to mix the exact chemicals and elements together causing a massive explosion setting forth the evolution to any and all life he would have had to have a large beaker. Any change to environment would have conditioned genetics; for instance there is a deer called the short leger the reason this deer has short legs is its terrain is steep mountains which likely was relative to the fact that going uphill was less distance to the ground conditioned motor memory to nuero transmitters making genetic adjustments? And that is what it's like when you become a physicist asking any question to how anything evolves or exists there is actually no way to know many things which baffle the minds of science. Religion asks the question why, there is no why as why assumes there is any cause there is no cause there is no authority there is only casualness, science asks the question how' where there can be any comprehension to anything factual anything factual can be realized where there can be any dialog through language; Human language originates from the comprehension and interpretation of cultural relationships attempting any communication as in the origins to technology where people began transferring forage occupation to our economic interdependence where there is an assumed responsibility, so assume the position to authoritarian obedience where everyone is expected to pull the weight of our material domestication also causes social antipathy or pretension which is stimulated by climbing up the flag pole of nationalism where the concepts of a collective union are symbolized in memorialization largely by the concepts to democracy and defending freedom, the concept of freedom postulates the constraints of tyrants and or authoritarian perversions of control assuming ownership of the people as slaves as any law infers there is no freedom of choice only compliance and conformity to a mysterious aura which conditions any sense of paranoia where the chains and shackles rummage in the auspicious mist of distrust to a paternal discord. The greatest question in youth is what is there or what does anyone know; however this is by and large conditioned by what education does postulating there is a common historic world view everyone knows in a bias idiosyncratic association maligned to the unsettled waves in the strife for progress in the wake of memorialized triumphs and defeats. The true world view is a maze of actions and confusions enters the political and economic doyens offering up any labyrinthine analysis in the yarns of defining solutions to the world's relentless conflicts conditioned to the arena of theories and ideological assertions inflamed largely by religious speculations.

Compliments lure any pose in any measure to successes anchors identity in mirrored reflection postulating contribution in attraction

There is no cause only casualness in adaptation to the physics of our culture

A consumed mind conditions stress which is a constant in the pressures of life's expectations tied to a leash on a treadmill chasing a future time in self-sustaining prosperity stimulating the imagination and consummations of desirous dreams and fulfillments, now how about a little S & M with a woman with a billed black cap and leather whip and her name was Desire. What does define perversion assuming any moral authority is relative to the definition of criminal, criminal can only be defined as any intent or action to harm another individual; lawyers would have nothing to do if they did not invent more laws, there is no science to law, laws are a necessary deterrent due to the perversions of our culture and capitalism where anything can be turned into a law license taxation or business as the origins to economy are physical mediums in transfer challenges any conceptions to democracy. Democracy assumes a collective cause is known however there is no true constitution providing any interpretation to what conditions human existence which is what technology does as the driving force to our culture. Technology conditions time in the perpetuity of changing fashions like watching Back to the Future, however there is no physical space or time there is no physical past or future only the moment where all motion and change exists. Language is always a half step beyond experience as the mind is often looking backwards in speculation signifying any realization where most experience has no reflection disseminating relativity.

Our culture has a force which stimulates a resistance enters an anarchistic polarity looking over the shoulder checking approval which is relative to a paternal malady where human emotions are conditioned by many factor however conditioned primarily to an inability to communicate many common experiences which are communicated by symbols rather than any true intuit. Existential dualism is seeking approval wanting to be known for being successful making any contribution.

The question is why the debate, the answer is why has there always been so much conflict to any assumption of authority providing any direction, the truth is that direction is conditioned by what technology does which has been ancestral relative to the underlying current of technologies providing innovation new and improved modern conveniences which also expands economic positions accommodating an increasing population calls for better education to fill these positions so there is better placement to have a better functioning society since we are held to a moral standard conditioned to authoritarian law which is actually religion conditioned to a paternal sense to order, now read the bible metaphorically and you then may become a philosopher attempting to provide any interpretation to our culture anyone could comprehend, the main theme to the bible is cultural anthropology however written by an artist who wrote the bible in allegoric metaphor in the question of quantum theory, where did anything have any origins what does anything do and where does anything go stimulates theoretical science. There is no such thing as an intelligent design to the universe there is no such thing as physical time or space there is no past or future existence there is only the moment of all action; there was no human knowledge before the evolution of the human culture without material culture there would be no language' any knowledge evolved from any interpretation to relationships, there is nothing known which has never had any experience there is no pre-destiny, there is nothing true reading any books where there is no comprehension to any factual existence.

The theory to quantum physics is relative to conscious experience and the way we perceive reality which enlists any relativity in perception to experience in life's adaptation to cultural domestication's and the perpetuity to technological innovation stimulating current fashions and accessories evolves language reality is the linguistics to our existence. The problem with science and theories is relative to any context asking the question in defining any physical existence.

A theoretical psychology in diagnosis to the geneses to Autism, The individual "I am" perceptive degree to awareness" in infancy conditioned in limitation to physical coordination and mobility has a sense of paralysis conditioned by muscle tone and hand eye coordination precedents challenges to an instinctive nature of assimilation to environment with frustrations conditions polarity to dependence of instruction where there is no intuitive knowledge to culture where perception is subject to memory rather than practicing construal to relativity as an individual where there is no pre conceptual operativeness to consciousness in adaptation.

Sheldon' Big Bang Theory; Autism is an acute inquisitive nature to perfectionism disseminating perceptive associations to reason gravitates to logical assessments' Construal requires contextual construct in comprehension to cultural adaptation in early development patterns to memorization are relative to soundex associations to language development in orientations to ornamentations to physical mediums in the material epitome has significance in developing comprehensive relativity to culture in self-dialog knowing what you know making sense to involves.

Thought is suppositional speculation' thinking construes contextual relativities

Education programs idolizationin memorialization to cultures emblematic monuments

Smart people ask the question what education is telling me to know' assumes a common perspective

Metaphysics is poetical language of human cultural

In most education there is no practice to routine applications automating memory recall, any intuit recall has any contextually prompted relativities in comprehension

Autism would be relative to adaptation to a preconditioned material culture in observance challenging any cognitive ability to characterize any association to sounds, tones, mannerisms, physical sensations, ornaments and embellishments. The origins to language are any adaptation to our material culture soundex in association to identity and relativity. Origins to any indigenous language are relative to the verbalization of experiences defining relations in self-dialog and anything communicable within the transcendence to cultural heredity. Erratic behavior and aggression would be relative to an over active mind lacking any definitive structure in achievement conditioned by a domesticated compliance in association to an over bearing nature to codependence.

Baby talk is non intuitive. Autism is misdiagnosed due to the power of suggestion in the assumption there is any genetic defect can affect parental disposition where any sense to imperfection can condition a paternal dissonance. Labeling children with autism and cousin ADHD causes a peripheral stigma postulating any genetic defect conditions subconscious objectivity. Any parental anguish is relative to any parent wanting children to be intelligent where expectation has any conditioning to perfectionism.

The force in what you do in adaptation to tradition has no art

Children have an overwhelming desire to do things on their own being independent wanting approval for being successful where ornaments such as certain toys stimulate coordination and functions to outcome like a Xylophone where coordination develops motor memory to rhythm. Existential dualism assuming a world view is a parental projection in the assumption there is a motion picture to time where culture changes, people have little to no recollection of their own perceptions to reality when they were children-youths relative to the emotions to social antipathies concerning inhibitions surrounding malicious behaviors. There is the physical world and the metaphysical experience where perception is conditioned by any knowledge to infrastructure and social fraternity.

The brain is the processor of Nuero transmitters (the nervous system) of physical sensations providing motor functions; separate the mind in observance is stimulated by the sensations of visual sound touch smell interaction and emotion. Stress is conditioned by the antipathies to life-sustainability combined with the consistency to the physical sensations of clothing conditions awareness. Motor function and coordination surely can be a factor in delayed development connected to any sensory neurological continuity in awareness coordination to environmental chameleon adaptation.

In western culture the main confusion with education is relative to individual application or finding one's own specialized field, it's not what you know it is how you know anything without being scripted to memorization; most education has no intuitive construct in cultural relativity to social and economic function.

In 1945 at the end of **WWII** was the start to the baby boom which led up to the hippy revolution which was caused by over population and a deep recession rousing a rebellion against authoritarian rule wanting to change the world to a communal way of life.

Countries have no principles people do

A collective conscious view is a memorialized impression

Construal is relativity to functions including culture is the code to comprehension

There is no true world view

The truth is there is no cause there is no authority there is only authority in knowledge to culture

Contextual relativity conditions comprehension

Fame has a false presence to knowledge

Technologies obsolesce craft

Conformity assumes a patriarch collective motion picture to time

One may think oneself insane questioning what intelligence is where there is no instruction to culture

Knowledge in philosophic interpretation provides confirmations to common experiences defining cause

Procreation and wanting a life partner is an ardent desire in cultural fulfillment

Notes:

Section Thirteen

It is fatuous' people believing in or worshipping "Jesus" when in reality what is written in the bible tells us what we know... culture is what materialism does makes estranging tenants piling up in a bemused quarters malady... so it doesn't make sense someone would crucify a poet for simply wanting to educate others on the nature of our culture and that anyone would stand by their rulers who assassinate poets or philosophers whom make sense like Socrates... If Jesus somehow saved people why would they have stoned him carrying his cross and why after his resurrection and departure would so many people come together in a movement to change what? If Paul or Jesus had been so influential in changing the world *(as impressioned by the calendar ad/bc)* where are the statues as there is of Socrates who lived and did die at the will of the Romans and where are any dates of births as there are for Homer, Socrates & Plato. The thing is that nothing ever changed being that the basic nature to our culture's structure is innovation and consumerism and it is simply true that there is no collective mind only what media's impersonate postulating one mind. Another question is relative to perception and states of mind which of course are conditioned by many facets and impressions to time so how can or would anyone know another's perception or their world view postulating a collective conscious mind. Interdependence has an expectation is the Adam Eve parable conditioning world views in reflection wanting to be successful.

Quantum Physics' Existential Dualism "They Became Aware" there is no collective consciousness there is no collective state of mind is the Adam Eve paradox, material culture is interdependent in forage transfer there is no "God "collective mind to knowledge' there is only metaphysical science to cultural interpretation

Quantum Physics' Culture impressions consciousness in adaptation to forage transfer conditions interdependence in expectation to natures forage in agricultural despotism

Before material culture the Garden of Eden had no expectation

Context in biblical comprehension is cultural relativity which the "teachings" mentioned by Jesus are written in parable…

The main theme to the bible is one mind knowing our culture there is no virtual knowledge to anyone's life

Sensing dejection postulates exclusion in value conditions chameleon dispositions

Pretending to know anything postulates any intuitive relativity

Irritation has a vulgar disposition

You feel awkward when you use a word you don't know

There is no other blind conviction in faith that divides people more than religion

The following is poetic metaphor for non-virtuous or an unprincipled persona… There are consequences to certain behaviors conditioning personalities and karma. The other aspect to belief and moral is relative to the ideology that there would be two places sorting noble and malevolent * heaven verses hell used as an ultimatum provokes temperaments to expectations conditions pressures to perfectionism in a world laced with trap doors tenured with concierge emancipatory paternal wardens.

In the Sermon on the Mount Mathew VII 26-27 And every one that heareth these sayings of mine, and doeth them not, shall be likened unto a foolish man, which build his house upon the sand, and when the rain descended and the floods came and the winds blew and beat upon that house and it fell and greater was the fall of it…

Was King James Version" of Plato and Socrates, like the bible philosophy distinguishes the same characteristics to our culture, "there is no virtual knowledge" to the life of Socrates only anything written down including Plato's experience which can be related as Paul to Jesus is too significant to ignore knowing what knowledge provides in confirmations modifying memorialized perceptions. Virtue is built upon knowledge to our culture knowing what anything does.

There is no virtual knowledge to the life of anyone: no one knows any others reality tunnel in perception as movies do impersonate in the opposite as there is no comprehension within what anyone says regarding another's experience in stories innuendo or gossip setting any precedence in any postulated characterizations.

Mark IV: 11-13 and he said unto them, unto you it is given to know the mystery of the kingdom of God: but unto them that are without, all these things are done in parables: That seeing they may see, and not perceive: and hearing they may hear, and not understand; lest at any time they should be converted, and their sins should be forgiven them. And he said unto them, Know ye not this parable and how then will ye knoweth all parables? Context

The term sin is relative to any iniquity to what one does to oneself in misdemeanor or intentional contravention setting precedence in self-indignation affecting any memorialization

One Step Closer; Linkin Park

No one knows any others virtual reality time line in experience perceptions or dispositions

Authoritarian nose is an oppressive intrusion perplexing intuitive retort

The term Christ is noted as a paternal being "Our Father" we are all paternalistic; paternal is domesticated intuit, IE law conditions a paternal sense to order to preserve liberty of mind

Twelve step meetings are an intellectual practice in comprehension so what was Jesus Christ's intellectual venue when intellect can be inhibiting unless there is a content setting

Mathew V: 38-41 Ye have heard that it hath been said, An eye for an eye a tooth for a tooth: But I say unto you, That ye resist not evil: but whosoever shall smite thee on thy right cheek, turn to him the other also. And if any man will sue thee at the law, and take away thy coat, let him have thy cloak also. And whosoever shall compel thee to go a mile, go with twain

Mathew V:38-41 parables the roots to transgression is conditioned by social manner in any degradation of shame associated with cowardice/machismo and the shield of virtue indifference as intuit knows has a consuming affect and conditions reprisal and vengeance in memorialization is the law of the people and in authoritarian transfer also draws the authorities of law

Islamic Sharia law is primitive to an indigenous heritage unevolved to modern society's fashions having similar paternal distortions and infirmities

The Movie "Naked" Mike Leigh is an accurate depiction to social transgressions raw nature

The only psychology educated during the seventies in teenagers asking the question' *"what is psychology I don't want to be stupid...*" was Freud…The Ego and the Id… it is ironic that today due to the internet there are millions and millions of people sharing things in writing or in videos, some enlightening some more confusing however it is out there now since people today due to technology have that means which did not exist in the past… It is also true that reading posts while analyzing content to respond or retort practices philosophic hermeneutics as intellectual dialog can be intimidating. It is also true that dialogs have improved significantly over the past ten plus years since a revival began post 911 which still have always existed in intellectual forum. The point to philosophy is to make it easier for people to practice comprehension of things shared in common to our culture. Philosophy also provides confirmations to common influences which stifle mediations.

If there was anything to Jesus there would have been "written" common knowledge changing the way people think, though truthfully childhood is conditioned by cultural inhibitions in the passage to adolescence, consensual age and independence which is simply what people go through to become conscious beings to our culture.

There is no other record to a one mind movement… other than the 1960's rebellion "which alarmed authoritarians" however there was no collective venue other than social gatherings, arts and civil rights protests against despotism as philosophy then had limited knowledge… (*Read Acts after Jesus left in a cloud people gathered and the numbers multiplied*) The fact is that people do not control the culture the culture controls the people as technologies evolve changing and expanding economies, which has a consuming factor where there is no comprehension to provisional change which today is substantial within the many options to occupations truly perplexes education and government administrations.

There is no collective conscious mind there is no collective mind of knowledge keeping track of anything... The only thing that can and does last would be what is written down in books, and that is the comportment of the philosopher... Before the Old Testament was written there were limited books on history IE the printing press wasn't invented until 1430. When books prior to this would have been scribed individually and only available to certain people of wealth and religious organizations... There was no recorded history being kept from any origins... So this is likely why the bibles author wrote an allegory starting with Adam and Eve creating a genealogy that did not exist, this simply a practice in quantum theory. The author then writing the New Testament based upon Plato and Socrates which is filled with metaphors expressing social subjugation and existentialism... Hollywood has been involved in one mind art since the 1930's however there is/was never such a thing as one mind or a collective resolve as the calendar impressions assuming a new beginning. Allegedly (Wikipedia King James Biography) the bible was not even published until 1611 by King James who is said to have converted the Roman Catholics to the gospels of Jesus Christ; the question is; how does a person's existence save people from adaptation? The truth is that knowledge does have an estranging affect in what we think changing any perceptions. The truth is that comprehension of anything in existence requires context... all else that we know are simply memorization's, the fact is that the bibles metaphors are extremely difficult to comprehend while applying to what you know or have experienced. We all have knowledge by experience in culture.

God is knowledge is best expressed by the biblical character Paul not Jesus, Paul is written in first person however' the bible claims several authors which doesn't make sense anyone could have followed Paul around writing everything a person says or experiences with any accuracy. There were no recording devices but then remember allegory supplies the context for metaphors aimed at the reader to connect with... It is also true that the bible allegories transgression which is what we go through to be conscious beings conditioned to human ingenuity in evolution; without technology we would be primates.

Plato's works can be read at the following websites

http://classics.mit.edu/Plato/meno.html

http://classics.mit.edu/Plato/seventh_letter.html

http://classics.mit.edu/Plato/republic.8.vii.html

http://classics.mit.edu/Plato/apology.html

The bible allegories despotism which is triggered by economic recession and depression waking people up aspiring intellectual revolutions which can also be seen as how human intelligence has been evolving from a past that was always more harsh and unforgiving reasoning that economic infrastructures have evolved by innovations in efficiency to meet the demands of an expanding culture providing positions for fluctuations in population which is more embryotic in the nature of transfers to capitalism in human mediums which condition interdependence: "in a time where science was heresy" the story of Paul and Jesus can be compared(though mystified with magic) to the dialog and views of Plato and Socrates which has been documented in Plato's writings Meno, Seventh Letter, The Republic and The Apology "though mostly Meno" and it is true that Plato's writing shares the same passion in the comportment of the philosopher providing cultural interpretation in relation to society and governance to virtues. IE "Sharing similarities to Paul" Plato's many Letters he called Epistles, The Republic and The Cave which is relative to quantum physics reality tunnels in the perpetuity of change to "existential" western culture fashioned by innovations in modernization. The main theme to One Mind is relative to understanding the nature of virtue which is circumstantial indication the author used Plato's writing as a guide, much as have been used by this author using other's notes most inspirationally Georg Gadamer. One mind is mentioned by Paul at the end of Corinthians II, and is also expressed in Act's "They shared all things in common" likened to the Communal life movements of the 1960's.

The letters by Plato may shed light on the epistles of Paul

The real thing is that since there is no one who saves us from social distortion there must be factors to be known in the causes of things such as homosexuality... Such as carnal compulsions stimulated by sexual arousal... It's simply psychology which is what is metaphoric in the bible though who reads it when people don't know how to apply metaphors to common infirmities. Young people going through puberty today are being influenced in the likes of celebrities and any law suggesting anything un natural is acceptable simply because there is a law and by what moral authority makes anything conventional... and actually religion is pretty much based upon faith and worship which is opposite to the actual philosophic education in the writings of Jesus or Paul which people simply do not know how to comprehend reasons why there seems to be more and more confusion to social degeneracy. The time for philosophy is youth and then for those in recovery from addiction depression or emotional infirmities. Authoritarian philosophy is obedience however lacks the definitions to intuitive assessments of the influences to rebellious characteristics in dissention convening any dialog to accordance.

Providing expansion to economic positions the original source is human physical mediums in transfer example' "Technologies" robotics... replaces labor, Cultivation is the origins to civilizations settlements evolving crafts modernizing convenience's evolving from barter to money and capitalist economics in forage propentence transfer.

What is inalienable conditioned by interdependence?

Consuming, most conversation has no intuitive juncture' the student is the teacher, dialog is best posed as a question, what is the question' you have the answer you just don't know it, Individual interpretation is a dyslectic menagerie I Corinthians XIV-1-14 the social malady.

Democracy' if you read the New Testament properly you would know the true Christianity would be a twelve step meeting; Social Venue meetings could be sister'd into 12 step charter' a place to converse' network and mingle...

Before technologies in cultivation as forage gatherers humans lived as nomadic clans following harvest being part of the food chain.

Culture challenging intuit/comprehension is a consuming factor is the Cain metaphor, Genesis Chap 4: 1-16

Consumer becomes consumed...~Marshall McLuhan...

There is no comprehension to our culture in expectation to being successful... leads to obsessions to escape in alcohol or drug abuse, consumption limits comprehensions using construal reasoning cultural relativities to individual or collective applications... Is another way to interpret the Adam & Eve metaphor to cultural adaptation.. Providing confirmations to common infirmities 12 step meetings are good sources for community though have limited

resources describing the physics to our material culture…

Metaphysics is science. Religion is dogma.

There is no collective mind to knowledge assuming an authority is the metaphor to Adam and Eve wanting approval in the invention to an intelligent design is the origins to existential dualism or interdependence.

The previous is the context to reading the bible… If you do not know this parable how will you know any others…~Jesus? Or metaphysical science describing how the mind is conditioned and challenged in adaptation to our culture of ingenuity

Trouble shooting anything requires physical relativities in construal develops mechanical skill in the practice to comprehensions includes the artificial intelligence of the means and any social relativity. Mente Concipio; ~Galileo' I worked out experiments in the mind before performing them physically.

Alzheimer disease is relative to consumption in the practice to cognitive skills in personal manageability in self-sustainability also conditioned to age and a sense in personal value or contribution

The first step to recovery is humility

God grant me the serenity to accept the things I cannot change the courage to change the things I can and the wisdom to know the difference.

There is no collective conscious union of mind... There is no intelligent design knowing cause it is not what you know it is how you know anything is any practice to comprehension

Another Brick in the Wall... Pink Floyd

https://www.youtube.com/watch?v=m55RDNlWnLI

Being there is no intelligent design there is no preceptual operative nature to conscious experience there is no "One Mind" there is no collective realization to cause however paternal is intelligence as conscious beings.

Meth is the most consuming drug in society which accentuates a desire to be constructive which conditions mindless acts trying to keep busy has a degrading effect on reverence conditioned by a social sense often lends to seclusion assuming anyone knows your perception

Social settings can be very confusing depending upon one's own involves where often there is conflicting ideology relative to cosmological postulations contesting intuit.

Dyslexia is relative to the physics to language use in relativity to culture defines any selective literacy

If you experienced the trauma of social dysfunction in childhood thinking later in life the world had changed the same exists today as it always was due to "mainly" poverty, so how did anyone die saving people from the sins of greed in capitalism.

A three legged dog knows no boundary

Social distortion is any depth to knowing any malady in who knows what of who knows anything

Philosophic Hermeneutics

Adam knew Eve and begat Cain? And who did Cain marry when he left to live in the land of Nod... seriously if you do not know this parable how will you know any others

Mark IV

11 And he said unto them, unto you it is given to know the mystery of the kingdom of god, but unto them that are without, all these things will be done in parables;

11.5 What state of mind knows any cause? Parables are poetic summons

12 That seeing they may see, and not perceive; and hearing they may hear and not understand lest at any time they shall be converted; and their sins should be forgiven them

12.5 The term sin is a suggestive assumption relative to intuitive comprehension to self-knowledge in adversity to cultural oppressions

13 And he said unto them, Know ye not this parable? And how then will ye know all parables;

13.5 Context to knowing what you know in experience to culture impressions

14 The sower soweth the word;

14.5 Construal, We all have knowledge in transgression to cultural adaptation in limited comprehension, is the cause to philosophy

21 And he said unto them, is a candle brought to be put under a bushel, or a bed? And not to be set on a candle stick

21.5 Context is essential in comprehension to any experience

22 For there is nothing hid which shall not be manifested; neither was anything kept secret but it should come a broad

22.5 We all have any knowledge in cultural occurrence

23 If any man have ears to hear, let him hear

23.5 "opened mindedness" humility

24 And he saith unto them, take heed what ye hear; and what measure ye mete, it shall be measured to you; and unto you that hear, shall more be given

24.5 Knowledge is accumulative and progressive knowing what you know knowing anything

25 For he that hath, to him shall be given; and he that hath not, from him shall be taken even that which he hath

26 And he said, so is the kingdom of god, as if a man should cast seed into the ground

27 And should sleep and rise night and day and the seed should spring and grow up he knoweth not how

28 For the earth bringeth forth fruit of her-self; first the blade, then the ear, after that the full corn in the ear

29 But when the fruit is || brought forth, immediately he putteth in the sickle, because the harvest is come;

29.5 It is difficult to know anything absolute/drawing conclusions assuming any dominion

30 And he said, whereunto shall we liken the kingdom of god? Or with what comparison shall we compare it?

30.5 What are any determinations defining virtues, no one knows how anyone knows the world contests intuit

Mathew VI

22 the light of the body is the eye if therefore thine eye be single thy whole body shall be full of light

23 But if thine eye be evil, thy whole body shall be full of darkness. If therefore, the light that is in thee be darkness, how great is that darkness!

24 No man can serve two masters; for either he will love the one and hate the other; or else he will hold to the one and despise the other ye cannot serve god in mammon;

24.5 God is a state of mind; being one mind is being centered'

Mathew VII

1 Judge not that ye be not judged,

2 for with what judgment ye judge, shall be judged, and with what measure ye mete it shall be measured to you again.

3 And why beholdest thou the mote that is in thy brother eye, but considerest not the beam that is in thine own eye

4 Or how wilt thou say to thy brother, let me pull out the mote out of thine eye; and, behold, a beam is in thine own eye?

5 Thou hypocrite! First cast out the beam out of thine own eye; and then thou shalt see clearly

6 Give not that which is holy unto the dogs; neither cast ye your pearls before swine, lest they trample them under their feet, and turn again and rend you.

7 Ask, and it shall be given you; seek and ye shall find; knock and it shall be opened unto you.

8 For every one that asketh, receiveth; and he that seeketh, findeth; and to him that knocketh, it shall be opened.

8.5 There is no purpose there is nothing to know, it's not what you know it is how you know anything, consciousness is conditioned by our interdependent material culture…

Notes:

Section Fourteen

We all have knowledge by experience, God is knowledge' god is not the author of confusion...

The whole concept to gods is more relevant to understanding existence postulating an intelligent design IE; what defines knowledge in reason

Consensual age is an ambiguous virtue

Generalizations are a fool's pardon

Knowledge is relative to what conditions human experience

Context is the most important aspect to any comprehension

No one knows who you know only how you are

"Existential" is the physical embodiment to the material culture epitome

Subconscious mind is a cultural directory to experiences

Thought is speculation thinking construes intuitive realizations exercising any comprehension

Human perception is conditioned by a material epitome impressions memory

Nothing changes anything one does

There is no physical collective conscious mind only artifacts and monuments

There is no motion picture to time

Time is a condition to a means, there is nothing or anyone keeping track of everything

Environment is a familiar idiosyncratic impression

There is no true motion picture to time IE a world view' is what version

Memory has any selective slide show to past experiences of significance

Marriage merges world views to any assumption of times impressions

The whole planet would have been more tropical during the age of dinosaurs

Did humans evolve from little people, is a curious supposition

When you do anything degrading to yourself or anyone there is a sinking feeling in remorse

Context is the essential component construing anything in speculation knowing anything true

No one knows how anyone is known by anyone is any idiosyncratic polarity

Dysfunction in family has a posterious aura

Communicating intuitions are often guarded by paternal inhibitions

Notes:

Section Fifteen

Desiring material adornments is a slave's parlor

Religious faith pretends to know anything

Children being told what to know or how to do anything contests competence

Memory is impressioned by pendular significance

Human interdependence is what culture does is conscious objective contesting intuit

Stupid is what stupid does

Human medium transfers in convenience have a reverse in requirement

Authoritarian police state is the worst oppression incites rebellion

In puberty; erections "*painful*" cause a temporal fixation and ravenous carnal compulsion conditioned by social influences is likely a key to abnormal sexual orientation

Economic circumstances estrange supporting obligation in divorce knows any anguish

Dominion is a posterious circle postulating eminence pillories inclusion

During formative years social subjugation is a mischievous zoo

In family the social view is a paternal association

In adolescence there is a different sense of community

You want to go to your high school class reunion to be what you were not

Adversity should always be seen as opportunity

The question of religion and science is' where does anything go

Crazy is thinking there is anything to know

An adolescent world view is an artisan fashion

Nothing is as it ever was anyone knows

A social view is anyone's own asylum director

A "mis-demeanor" ordinance' competence implies a unified code is a conformed view

There is no cause there is nothing to know it is not what you know it is how you know anything if you do not know what you do not know you will not be foolish

---

Conscious beings evolved from single cell organism volution requires a lot of thought in any attempt to understand any physiological occurrences

Transgression is the adaptation to a material domestication in expectation

Notes:

Section Sixteen

A world view is an isotopic spectator

Adolescence postulates independence is consensual age is shorted lived

Youth postulates a maturity station to adulthood

Desperation affects paradigm transcendence

What you know and what you say often has two layers wanting to be right

Adolescence knows a virtuous adversary

Memorialization conditions rhetorical thought in idiosyncratic critique

20's are an empirical age

Godhead 'God is a state of mind, virtue knows any intuitive value

Memory is always looking back projecting forward wanting to know the story

Manipulation deception consequences

Subconscious is the hard drive processor of cumulative experience

Money conditions a permanent temporal complex

Military uses a form of dejection in boot camp training breaking down ego into authoritarian submission

A Rebel without a Cause; James Dean; Male aggression in groups challenging courage seeds transgression

Women want a secure male also provisos financial sanctuary in trade's occupation

Corporate national news is the government agency

All sexual abuse is relative to a carnal compulsion stimulating male aggression; carnal compulsion is caused by arousal stimulating a temporal fixation on "release" male aggression is also conditioned by "female dejection"

If you are over consumed by pressures from a hard week of work relaxing with alcohol may not be the best remedy

Adam & Eve "became aware" conditioned by the existential material culture in vanity

Girls want fathers as good role models for future husbands

Time is pressure

Thinking is an intuitive director

Every country has its own idiosyncratic polarity

Notes:

Section Seventeen

Reality is what you know

Wanting to be successful conditions egocentricity

Conscious polarity is a paternal muse

Organizational skills keep any mind constructive

Rebellion knows an emancipatory cause to an authoritarian dictorial

Tally Ho

Martyrdom has a loathing appraisal in social valuation

The force in what we do to live lessens any art to adventure

A world view is one's own reflection of time

Schooling environments authoritarian regimentation

In school social antagonism armors law as an authoritarian defender

Education impressions our view to a cultural kaleidoscopic memorialization

Most spelling is memorized

Technological culture conditions interdependence and conscious reality

Science is the key to the mysteries of life

Addiction is the physical compulsion to escape the material yoke

Lethargy' grumpy modes trigger hostile responses

The transfer in authority to government directs a predominant assessment

Indifference knows no hold

Paternalism is a maladroit affiliation

There is no collective mind only individual perspectives and any scientific fact

Any world view is an economic position

Economy is a maladroit sibling

Transgression is a disease of social & economic discrimination

Playing any sympathy cards is a fool's hand

Notes:

Section Eighteen

Resigning from the establishment is an anarchistic notion

Overzealous natures end in perversions

The real world is isolation for most people

Possession is a carnal emotion

Meditation 'dualism is the existential yoke

Philosophy is the science of mind to culture

Perception is an intuitive assessment

"God is not the author of confusion'" is written in Corinthians I XIV, much thought has no intuitive juncture. It's obvious the bible was written by a poet

There are more opportunities for men

Stay out of other people's business and you won't be befuddled

The social panoramic view is a paternal orientation

No one knows who you know

But I suffer not a woman to teach, nor to usurp authority over the man, but to be in silence. 1timothy 2:12

Woman feign paternal dominion in heredity to domesticated roles postulating indebtedness

Women are more inclined to arc angels as mystical guardians

Postulating paternal supremacy is an ardent warden

God assumes a moral authority in martyring compliance

Metaphysics; you have to apply anything to experience to know anything

Children live in a mystical dimension of perception

Hollywood impressions sophistications which do not exist

Authoritarianism is a bipolar director

Existential dualism is the dependence to human physical medium transfers in economic domestication

Existential is the primacy to human culture

Authoritarian justice is no more than truth corrupted by another perversion

Cultural prejudices are an ambiguous parody

National pride is an emblematic fraternity

Notes:

Section Nineteen

Existentialism is an idiosyncratic wardrobe

The world is a glass menagerie in reflections to shattered dreams

Religious postulation is pseudo psychology asking why not how

Opinion polls are political persuasions

Hindsight distinguishes any honorable discharge to transient causes

Cultural perversion perplexes propriety to mind

Consciousness is one's own true view

National media conditions global social focus

There is no intelligent design there is no cause only casualness there is no one mind directive

Life experience is like a puzzle wanting to place all the pieces together attempting to know any reason

The' "Serious" clown

Desiring to be successful has a truant officer

Being wrong about anything has a left temporal inclination

A movie camera is an extension of social perception

Addiction is a material force in adaptation

Language is a provisional director

You have nothing to lose when no one expects you to win

High school is a happenstance menagerie

Music impressions social mind ambiances

Coordination in adaptation to a material culture conditions conscious construal

Perfectionism is a paternal dissonance

A sense of shame has a sinking blush

The authoritarian shield is a posterious defender

Personal manageability has a maladroit economic guardian

There is no normal culture

Notes:

Section Twenty

Transgression is any origins to law

Transgression is a force in adaptation to social and economic cultures knowing no cause

Rural towns have no attraction to corporations concerning the demographics to population and volume to consumer commodities.

Restitution limits memorialization

Addiction is a ferocious coinsurer

What you measure makes you who you know

Law is artificial intelligence has no intuitive juncture

Sizing someone up is a tailors suite

Time is a perceptive aura

Greed is a fool's bounty

Time keeps an eye on a motion

A social survey is a superficial spectrum

Light has no physical congruence, if there was only a sun in the middle of nothing there would be no illumination

What is it makes anyone who they are

There is no reason to anything; intelligence knows what you know...

Social class subjugation in high school is a maladroit association

Liberty is a state of mind not a national emblematic gesture

All things in moderation tames ego master

Women have a sense of value where there is any appreciation; is that feministic

Much television sexual content stimulates carnal emotions

Most television perversion caricatures ethical relativism

Censorship postulates ethical standards

There is no collective dialogue knows any cause there is no collective conscious mind

Most authoritarian opinions are non-intuitive knowing casualness

Social culture has economic partitions

Section Twenty One

Elicit behavior has a chameleon influence in provocation

Belief has no intuitive comprehension to mystical apparitions

Rhetoric knows no contextual cause to resolves

Cognitive psychology is relative to social economic manageability

Despair has a sinking feeling of helplessness desiring successes

Social subjugation has a sense to emancipatory causes

Reading stories comprehension is connotative memory

Perception is a cultural epitome

The social view is a paternal malady

Paternal senses stimulate carnal desires

A racing mind is a lab rat in a cage

Fred Flintstone is the average Stone Age modern man

There is no collective mind to knowledge keeping track of anything only what Medias impersonate as relevant

Cultural adaptation casts a spell in confusion assuming an authority

God is not the author of confusion god is an intuitive consult

Holy books corrupt philosophic interpretation with imaginary companions

Traumatic experiences introvert social polarity

The arts hold subliminal messages subconscious is a poetical condition

A police state mentality "authoritarian dictatorship/ communism" is a transient cause associated to social economic despotism is the trigger to civil unrest an revolution

Reality has any depth to cultural perception

Social comparatives are a derogatory directory triggers perversions in thought

There is no intelligent design there is no cause there is no authority

Define a cause all would agree to enters any moral paradigm

Culture conditions consciousness; we all have knowledge by experience in environmental adaptation with limitations to mediation

Nothing is as anything was anyone knew

Notes:

Section Twenty Two

Judicial in education assumes a guardian tutor

Women expect a man to take care of them is a consensual age paradoxical custodian

Conformity conditions regimented focus

What law does is transgression conflicts its emancipatory cause

Bipolar is a degenerative paternal dissonant resentment to an expectation

Sexual arousal stimulates a temporal fixation

Restitution assumes a debt to an illusory cause in equal accord

A temporal cause Just-ice is a reactionary assault in vengeance is transgression

Sexual promiscuity is getting too much press

Existential polarity; no one knows your world view

Religion knows nothing; intelligence is paternal

Words like "religion" have no intuitive comprehension, intelligence is paternal

Intelligence is' knowing only what you know

You don't know anything but what you know

Are you who you know

Procreation is procreation sex is a carnal compulsion conditioned by erotic predilection

Childhood naivety assigns innocence to adventure

Reality is our own illusion of time

Sanity knows no cause

Good conversation is a personable art

Economic destitution has a dark presence

Does what you have make you who you are

Problem solving is a contextual consultation

Keeping up with life is a consuming factor

What is is what does

Notes:

Section Twenty Three

Quantum theory is relative to human culture and the quantum mechanics to nature in any attempt to defining existence

No one knows anyone's internal polarity

Desire rouses temporal fixations in fantasy dream states

Self-deception knows Pinocchio's nose

The transfer of the law is a maladroit assumption

You remember things that didn't make sense

Sexual innuendo impressions carnal thoughts

Material domestication induces any proprietary possessions

Passive aggressive is a disgruntled slave

Empathy is a paternal seduction

Sexual experiences impression sensual imagery

Much speculation conditions random thought

Poverty is a democratic embarrassment

Temperament is an ambiguous social gauge in demeanor

Mis-demeanor assumes a certain social virtue say's the law…

Authoritarian rule has a pragmatic sense of union

Media impressions the modes in human conflict

The earth revolving has a centrifugal force

Life is mysterious when you don't know what's going on around us

Social perversion; you wouldn't know anything you were not impressioned

Assuming anyone can read your mind knows you is a posterious acquaintance

We often say things that don't make sense

There is too much pressure in the world

As a child it is a common fantasy to have magical powers

When you are young you have more time to practice

Authoritarian dictatorship at airport checkpoints is communism

Notes:

Section Twenty Four

The cause to the war in Vietnam's was to defeat communism

Age is very estranging

Children like to impersonate crafts wanting to do things on their own

Mathematics involves physical functions not logical assessment

Anything true is simple

It's not polite to tell people what to do unless you are the boss

The world is not so pretentious after high school

Why do you have to sing along with songs when you can simply listen

There is too much going on in the world to keep track of

When you are listening you are always anticipating any relativity

---

Confucius say' who know what is anything

Trust is an earned reward

Life is a strange duck

There is nothing to know there is no meaning to life

If you want to reduce the chance of colon cancer you will fast for three says on occasion

Cognitive postulates domesticated regimentation

Any concepts to god and religion postulates there is anyone there for you in sanctity

Parental paternal authority is an awkward association

Any concepts to freedom is conditioned by material sustainability a paternal yoke any lairs of tyrants

Is government a socialist parent postulating psychological credential

Existentialism you are what you know

Existentialism is true psychology conditions conscious subjectivity

Speech impediments are relative to apprehension to contextual comprehension and repetition

Philosophy provides confirmations to common inhibitions in cultural adaptations

Any secret to being successful is dedication and practice

Notes:

Section Twenty Five

Exhaustion effects lethargy challenges any patience

Media is an ambiguous periscope

Law is a maladroit paternal transfer

Sanity assumes there is such a thing as a common world view knowing cause

History has constant revision questions educations assertions

The world has many entangled spheres

You want to wrap the world around you because you are so naked and afraid

Love is two shadows dancing to be one

It's easy to respond to people who make sense

All knowledge emanates from experience

Memories are an inside outside view lacking animation

Reality assumes a true world view what is

Infidelity is often caused by dejection

There's no such thing as what you think only what you know

Awkward moments are often scripted by rehearsal

Thought is reactionary looking for any connections to subjectivity

Cursive counter occasions due to a lack in mediation

Negativity should always be turned into a positive knowing any causes

It's hard to describe what you know about people you know

No one was who you knew when someone changes in exchange to familiarity in patterns to behavioral encounters

Women are more insecure because they have less mechanical agility

Nothing ever made sense and her name was echo

With anyone in conversation we often do not say what we are actually thinking like a game of stratagems

Life is a menagerie of experience there is no permanence

Sexual desire has an emotional compulsion

Notes:

Section Twenty Six

Over protective natures summon assistances is dependency

Time is conditioned by the perpetuity of changing fashions

Time is conditioned to change in perception

Transgression has two kinds social and economic

Social subjugation can be like a carnival variety show

Sanity knows what you are involved in

Paternal senses condition emotional compulsions

What did people do before there was any requirement for money in forage propentence transfer

There is nothing to know it's not what you know it's how you know anything

A world view; no one knows what's going on only what one knows in motives

Suggesting that no one masturbates is like suggesting there is no water in the ocean

First introductions involve posterious character impressions

Trust commitment fidelity possession suspicion seduction doubt control dejection

Subjective experiences jostle the mind

Social subjugation has a resentment is a paternal affliction

Love is having a greater cause than yourself

Dualism; no one you know knows your life recording edits any protrusion

Not knowing what's going on around us machinates involvement

Advert insecurities are very inhibiting

Violence on television causes many emotional impulses

Disturbing experiences impression negative aura's

Much thought has no council to people places and things

Facetious remarks provoking characterization are best braised with tethlon

One word can change a sentence to a question mark

Identity is stimulated by character association

Section Twenty Seven

When you are dancing are you dancing with yourself

Social manipulations want to twist any mind in affiliation

Time is a mental compulsion

Mystical moments are when time stands by

Dejection flashes a sinking feeling pursuits any lure

Mind is any mediation to position in social economic structure

Character is any chameleon gesture always in navigation tethered to familiarity

When you are a child you are who you are challenged to become someone you don't know

A child in the moment knows no tether

Time congests mind

Lyrics define the score

There is no collective world view recording

When you say stupid things that don't make sense you don't know who you are

In the primacy to intrinsic art nothing is original

Any concepts to god enlist any moral valuation and theoretic version hypothesizing creationism

A collective world view is one's own impression

Lust is any emotional attachment love has intuitive connections

Infatuation is an insatiable desire to share a union in mindful spirit

If there were such a thing as god how would it keep track of so many people god would be insane

Retraction in thought assuming any posture is any reverse to ego

Sticking ones nose into other business not knowing any facts is a fool's menagerie

Hostel language and verbal abuse trigger social antipathy

You have to have money to go anywhere

Anguish has a sinking feeling when anything falls apart

Losing control going into rage rouses contemplation

Section Twenty Eight

Awkward moments check orientation

Every new generation embraces its own idiosyncratic craze;
shake your booty in the carnal jungle

Famous people are subject to much public scrutiny in
example to ethical standards

Children are inquisitive sponges

Are you who you know in chameleon apparitions in
comparison

Compliments infer there was any imperfection postulates
dominion

Some things cannot be pronounced

Singing along to any song is always a little out of sync

Compliments condition expectation postulating perfection

Not living up to someone's expectations charms intrusion

Familiarity conditions behavioral responses

Jumping to conclusions knowing what you do not know is a hung jury in mind

Malicious is what malicious does

No one is born to any precondition only environmental influences

You are what you do

Defining genius; there is nothing known having not been practiced

Some words have no intuitive juncture requiring relative definition

Memorization is no appraisal to intelligence being scripted

There is no such thing as what you know there is no collective conscious mind

Science provides any physics for ingenuity and innovation challenges any mastery

Having any income is a most consuming factor

Social distortion is cavernous

Mathematics is the memorization to formulas and application to any physical relativity

Jealousy has any posterious incrimination

If you want anything too well in life it will always be evasive exasperating to be anything

Notes:

Section Twenty Nine

Uptight people make for sturdy book ends

Greed is the almighty oppressor money is god to many

Everything makes sense nothing makes sense anything makes sense

Exposing vulnerabilities caveats any manipulations to domination

Men's roles as the protector is very ostentatious; paternal law is deep and wide ranging

Keeping track of people postulating what's going on is very consuming

There is no such thing as a collective mind of knowledge; knowledge has two kinds what you know is going on and any relativity

Word processors obsolescence the art to hand writing

Who needs to know how to spell words where there is spellcheck

A positive outlook on life requires perseverance

Philosophy is like untying knots from a tangled fish line nothing makes any sense until you find the right orbit

Insane is ambiguous

Traumatic experiences can make a person insane in trepidation contemplating responsibility

Social influences can be very suggestive to paternal causes

Economic class segregation conditions social views

Doing what's right establishes any integrity

Personality distinguishes attraction in likeness

It's not what you say it's how you say anything resonates tone

Angry people challenge approach

Any debt to society infers any cause

Societal anxieties want to drag you down

Mood swings' you never know anyone until you have lived with them for awhile

Technology provides a sense of sophistication

Time alone is well spent in reflection

Adaptation to economic culture is very challenging

Notes:

Section Thirty

Stupid is what stupid does becoming what we do

Pessimism; Cynicism, distrust, doubt, glumness, negativity, nihilism, suspicion, hopelessness, pessimistic people are those whom have been unsuccessful fulfilling dreams

There are many positions to economy which aren't being filled due to matching qualification

Every economic position supports the whole of an embryotic interdependence

Meeting any social standards can be challenging where there is no continuity to expectation is an undercurrent to body language

Social humiliation has a stigmatizing affect

Economic position has a temperamental gauge

Structure has everything to do with having mind in focus

Authoritarian regimentation restricts freedom to expression

You do not know anything anyone tells you in second hand knowledge

Mainstream media impressions what to follow

Reason is the way of any truth

Nouns have no intuitive context limits memory to impressions

Any collective mind is a media impression

Trivial memorabilia assumes a view

Every generation has its own idiosyncratic culture

Any depth in perception knows any relativity

Speculation is intuition weighs any outcome

Practice in any art is time well spent

In puberty there is a natural curiosity in the innocence to exploration in sensuality often misguided

Naivety is a predator's prey

Drugs do not enlighten minds

Reason conquers any confusion

Authority in knowledge provides any contextual disposition

Helping other people when you can is a productive thing to

Notes:

Section Thirty One

There is truly no absolute democracy its free trade capitalism economy

The cultural sense is a collective economic opulence aspiring prosperity

News twenty four seven are sound bites postulating collective relevance

Government objectives condition collective views

There is no such thing as a physical collective conscious view

Money is the transfer in forage propentence transfer

If you are thinking about what you are doing you are not doing what you are thinking

There is no cause there is nothing to know

Derogatory epitome's swagger virtuous retorts

Social perversions are the main cause to religious convictions

Swank satirizes carnal seduction

Unsolicited advice challenges competence

Conquests for love knows a chameleon lure short lived

Authoritarian justice is a fool's guardian

Treason assumes noble causes

Excessive graphic violence in television plagiarizes social norm

Agricultural despotism capitalizes natural forage rights

There is no such thing as anyone being responsible for human culture

Homosexuality defies the nature to instinctive procreation

Edible cannabis has a creeping and lingering corporal affect

Authoritarian chastise has a humiliating orientation

Government statistics plagiarize demographic constants

Measuring what you do and what you know is a pendulum swing

Genuine affection has a certain appreciation

Paternal senses chameleon obligation

Notes:

Section Thirty Two

Sexual orgasms stimulate corporal ecstasy conditions a temporal compulsion in sensual desire

No one can make you complete don't fall for the first one that comes along

The Rocky Horror Picture Show isn't what you want to know at midnight

In origins to crafts early masons carved stones building shelters

Knowing no cause has a consuming factor

Authoritarian arrogance postulates ethics

Parental responsibility includes being involved with education especially reading

If you look funny to yourself become a comedian or an actor

If you want to be original with music practice your own rhythms

You are never alone with yourself in any art in what you do

Equality knows the same inhibitions

Television; viewer discretion is advised

Private property is one's own domain

Economic transgression is the perpetuity of change is a consuming factor

There is no true world view knowing cause

A collective consciousness is an internet carnival muse

Self-expression is inhibiting

Linkin Park is an intrinsic band

Most television sitcoms are lampoon exploitation

Every economic genre has any owns agenda

Ego is looking right over the shoulder wondering if anyone's watching

Memorizing nouns has no intuitive construct

Self-value wants to be a part to something constructive

AA Twelve step meetings are the closest thing to the true Christian method to being One Mind

The social view is a kaleidoscopic spectrum

Notes:

Section Thirty Three

When all else fails fall in a ditch

You're not crazy the world is stupid

Language is conditioned by frequency

Value in contribution has a monetary stigma

Relativism—Cultural Anthropology

Love is a paternal cradle

Acting experiences any role

Fame assumes a collective audience shares a view

Subjectivity in adaptation has no syllabus

Lying in bed knows a safe haven

Being with someone for life assumes a shared view

Contribution assumes social responsibility for any art

The bible is/was the most misinterpreted t art ever produced

Any world view is a cultural impression

Poverty knows a generic life

Life hands a dripping spoon is very deep

Dignity knows a warm bed

Philosophy is any mediation sorting out conflicts in edification to intuitive virtue

Nationalism promotes a superfluous story in an assumed collective mind

Abbott & Costello Who's on First (You-tube) I don't know now tell me

The ladder of success has intangible rungs

Nouns have any visual consults in physical associations

Movies were once more a stage venue transcending vaudeville

There is no collective mind to knowledge only any impressions

Homosexuality impressions minds in visual perversion

Notes:

Section Thirty Four

Transgressive offensives weaken equilibrium is a challenging experience to courage

Any state of mind knows any temporal condition

If you read well you will always have better comprehension

Movies stimulate strange thought assimilating roles

Are you going anywhere where are you going

Bad karma experiences dissipate as time in foggy dreams impressions memory

Suspicion conditions drifts in thought seeking associates

Any anticipation has a strange consult wanting resolve

Stuck in a rut knows no reward

Scripted knowledge has no intuitive visualization

Media speculation is an ostentatious peer

Pointing things out to people is an auspicious reflection

Bipolar is an inquisitive nature assuming a common sense

When someone doesn't like you they don't know you to well

If you don't know what you know you don't know what you know bout anything

Silver spoons earn no rewards

Authoritarianism is a posterious keeper

Social perversion is a subjective authoritarian clause

Be advised "adult content" is an authoritarian perversion

Familiarity conditions complacent modes

Cultural influences subject thought reasoning maneuvers

Super tramp Breakfast in America is intrinsic art

The bible has little good to say about people

Rhetoric; much experience challenges literacy in retort to corruption

Styx "Grand Illusion" assumes a collective cause

Notes:

Section Thirty Five

Time is a cultural condition

You have to have a job to earn a living nothing is free

Face value wears no blinders knows any liable

Indigenous languages may have limitations in cultural intuit

What is any head director? Doing stupid things to fit in?

Social essence intuitive communication is harmonious

Bipolar has relevance in material adaptation and intuitive virtues in retort

Personality is a mimes parody

What capitalism does in society in perversion certainties models to authoritarian rule

Liberal verses conservative is a political apparition

Socialism is a patriotic cause

There is no control over free trade capitalism if there was it would be totalitarian communism

The voice of the people is a mute choir

The ideological conflict in debate to a belief in god is relative to whether there is intelligent design or being any purpose defining authority

The collective view is a who's who social club postulating authorities' impostering accord

Wrong from right knows any remorse in retraction precedents behaviors

Restitution is apologetic retort

Predators are a fool's gamut

The heat of passion is a grandiose adventure

Cultural compliance conditions transgression in limitations to comprehension contextualizing cause

Humility weights balance

Know what you know when you know anything is a good practice

Intuitive social nuance beacons conversation

Yearning desire conditions patience

Any commitment has many impulsions in deceptive thoughts

Section Thirty Six

Arrogance knows a certain fool

Personality trumps appearance

Cultural literacy conditions perception

Innuendo is arbitrary conspiracy is everything

Routine keeps any mind in focus

Social speculation sights commonality

Denial pulls a maidens shade

Sensual desire is a most common fixation

Inappropriate remarks are a gestures joust

One mind knows the culture can never be fooled by anyone

Dejection instigates suspicion in an obscure territory

Sympathetic attention is an ostentatious lure

There's no such thing as a collective mind

Appearance affects prejudice in many applications scores equal opportunity

Occupations are accumulations of circumstantial practice

Opportunity has wide ranges to experience

There's no collective mind to knowledge

When students ask questions they are looking for "context"

Oppression is cavernous when your young and don't have much

Optimism drives ambition

Oppression knows material sustenance strife

The more you know about our culture the more you are one mind of your own…

Women want to be provided for

What is there to talk about when you are young wanting to be in love

Commitment weighs dowry

Notes:

Section Thirty Seven

Stupidity is competitive

Ever thought yourself insane thinking there was anything to know

There is no collective conscious mind knowing cause

Intelligence knows its culture

Government institutions have embryotic sources

Youth's generation's union knows a social apparition

Intuit postulates social norm

Domestic abuse is a discontent sibling

Thought is any pictogram to any speculation

Humility has a sound constituent

The social world view is a monetary pose

Having children appends another dimension to time

Democracy assumes a presiding obligation

Economic valuation establishes forage propentence dignity

Rudimentary education knows a scripted custodian

Human value knows a common class

Self-degradation tailors subterranean apparels

Positive sanctions reverse negative cavalries

Social personation has a deceptive lair

Civil gatherings beckon unified causes

Dysfunction in family emits a plagiary disposition

Sympathy lures a martyring cradle

Personal vendettas lure suspicious fortifications

A first kiss has a sensual interview to appreciation

Peering noses are fools beacons

Notes:

Section Thirty Eight

Isolation discerns isometric territories

During puberty masturbation is essential relieving carnal compulsions

During puberty wanting to procreate is instinctive conditioned by cultures consensual laws

Desire is a dream menagerie for perfection in domesticating marital tradition

Paternal intuit in conversation entertains a virtuous dowry

Economic adaptation initiates monetary subservience in material value comparison

Liberty is a natural right not to be owned by government; IE taxation is voluntary contracted by signature on W-9 forms

Family time in undivided attention is a good practice

Adam & Eve tree of knowledge; Make one wise' Cultural literacy is an intuitive schooling

Existential duality in physical mediums is true psychology

The force rather than any art to adaptation is a corporal enslavement

Family dysfunction limits social aptitude discerning embarrassing characteristics

Estranging relations charm truths to motives soliciting treaties

What are any origins to social animosity if not any sense in failure "Cain Metaphor?"

Success and failure in self-sufficiency is an economic forage surrogate

Global ideological conflicts challenge intuit (knowing culture) defining cause is relative to religious convictions to gender dominion

Culture in subsistence necessity conditions minds focus in authoritarian force paranoia in chemically altered emancipatory states

Exercise conditions a positive aptitude

Law suspicions character

An un-sober mind is a consumed mind to experiences where there was no retort

Intelligence "is knowing" what you know

Words require context applying definitions

A rebel with a cause seeks replies to an intuitive question why... wanting validation

How you know anyone is an idiosyncratic association

Defending honor? Is anyone trying to make you a man or a fool?

Notes:

Section Thirty Nine

Quantum Physics define human culture, consciousness is intuitive perception

God is you are, I am, I am you are God, God is intuitive perception

Government is not a private mentor

Government agencies are a beggar's pardon

Leisure conflicts with the pressures of reality

Any origins to life are cosmic biology "God"

Comedian caricature is a chameleon's measure

Are they holding you to someone or is that you buying

Environmental familiarity anchors posterious characters

"You've become a part of me" Linkin Park

Law is a convenient transgression

No one knows who anyone knows

There is no gain to knowledge using drugs it is acquired by experience

Before institutions social shunning was once a pilgrim's justice

Insane assumes a collective motion picture

Transgression casts a dazing spell

Courage fosters no leavers

Rhetorical distractions disguise ulterior motives

Pretended knowledge knows a trivial posture

Capitalism is a process of labels

Being conscientious is a paradigm pose

Daytime television is a fool's paradise

Agony of defeat checks value

Seduction knows a cuddling pout

How do you give yourself up to anyone in loyalty and trust

Notes:

Section Forty

Good company knows pleasure

Infatuation in loves desire is a tormenting experience

Male aggression is a transcendent instinct to domination in natural selection

As any sedative to stress having sex can have an addiction

Procreation is instinctive subject to a material dowry

In happenstance dejection has many barriers

Practicing any art a social pull conditions focus

Time wants a severance from occupation to leisure in sanctuary breaking chains at a gate

Gender acrimony strums a pendant harp

You want to go a little crazy to be sane

Using a thesaurus distinguishes definitions

If you never have any expectation from people you will never be disappointed

Freedom is the cause to Democracy what is it when free trade capitalism provides incentive overridden by corporate consolidations

There is no collective state of mind only what media impressions

You know anyone by character

What is living up to a social standard conditioned by law assuming a certain competence is offensive

Authoritarianism is certain arrogance

Charity knows social economic disproportions

Cowardly sells out fitting in being stupid

Authoritarian discipline for children is not intuitive

What you think and what you say to anyone is not always frank

The social view is an economic cultural impression in classifications "Got class"

Pretention has pinnacle inverses

In school popularity contestant is very divisional

Education impressions a collective conscious union knowing the same recording

Having a life partner challenges expectations in character consistency

Notes:

Section Forty One

Thought plays many suspicious roles in social exposure

Bipolar assumes a social norm; what is it

Gods, religion, authoritarian obedience has no intuitive juncture

Social dereliction is an animistic association in envious endeavor

RFID chips are setting precedence in access convenience is a techno doctrine

Corporate bias controls airwaves

Transgression is thinking oneself incompetent postulating there is anything to know conditions memorialization

Wanting to be popular has a sense of shame feeling unsuccessful

There is no intelligent design there is no predisposition to consciousness knowing cause

Consciousness is a cultural impression in the perpetuity of change in adaptation

---

Law postulates a social norm

Statutes to limitation" on crime does not make sense in sexual abuse cases

Law has no intuitive juncture

"Being Right" is a cultural challenge having knowledge

Boys having no structure or social involves can be mischievous wanting attention

Girls by domesticating tradition want to be provided for in exchange play the role of care givers

A man with no occupation has the worse trepidation

There is nothing worse than feeling like a fool buying any story

Having a comfortable partner in life has a certain confidence

Social dejection has a sense of mediocrity has a sinking feeling in isolation

Social manipulation is often ambitious to intuitive connections in any sense to humility knowing any same inhibitions

A sense of defeat is one's own voluntary oppression is that critical voice of mind in expectation

You know there is no such thing as a social congruent mind walking through a shopping mall

Assuming anyone can read your mind feigns any disclosure anyone knows your thoughts

Anxiety is always relative to confrontation where rehearsals in dream states only fruition where there is any reconciliation knowing there are echelons to circumstantial causes

Notes:

Section Forty Two

Perseverance knows a certain conviction to being successful plays many roles

Physical Graffiti by Led Zeppelin may be the greatest rock album ever created

In identity nouns are photographic memory in people places and things is not intuit, intuit is any relativity to function

Any obsession to alter a state of mind is relative to gratification and grandiosity living in a world filled with emotional snares

There is no collective state of mind other than authoritarian obedience postulating authority

Democracy assumes a role in responsibility what is it

Youth assumes an establishment in economic sustainability unknowing the relationship

Interdependence is conditioned by a transcending cultural disposition

Who anyone knows no one knows is reflective in the courses to commonality and diversity applies any impermanent distinction

Anyone can go nuts attempting to describe or interpret any experience or relationship challenging intuit where time is conditioned to utilitarian manageability

Television conditions thought to many perversions challenges any cause in freedom to expression postulating any distinctness to ethical standard where there is no intuitive dialog or intrinsic virtuosity

Vocabulary evolves in relation to economic medium function and progression to social ambiguity

There is no one mind conscious view only what media impressions postulating significance

One can believe such a thing as any god exists however no one has ever seen any god to know the truth as the debate between creation verses evolution persists John I 18 No man hath seen god at any time;

God can only be defined as intuitive knowledge to cultural orientation

Memory is conditioned by repetition anything practiced to precision conditions second nature motor memory like mastering a musical instrument

A physical medium of mobility a new car feels awkward until it becomes a physical extremity

Spousal abuse is agitated by incompatibilities relative to dominions to postulated supremacy habituated by insecurities and assertive retorts

Sexual relations have a sacred dominion in domesticated possession

Social satire has an ominous pose

What insightfulness wrote the dictionary

Trying to relate to certain lyrics can chameleon roles in exposure to another's experience

Linguistics are conditioned by social environment and economic profession

Embarrassment can lend to reactive behaviors in antipathy in any assumed exposure to uncertainties

The competitive nature to social posture challenges virtuous influence

Notes:

Section Forty Three

Geographic orientation is conditioned by familiarity impressions topographical perception

Egocentricity takes the center stage

The malevolence in much social media precedents facetious behavioral tolerance

Desire for the perfect mate stimulates fantasy dreamscapes

Homophobia is conditioned by traditional models to a masculine shield in sequestered intimacy

Reality assumes a collective movie knowing a script

Dejection trenches the mote where resistance may be the lure in a web to reign

Social polarity assumes any classification is normal to judge character reckoning any compatibility

Memorizing historic memorials in people places dates and events is scripted knows no intuitive construct conditions national dispositions

In school social distraction limits focus in dream states specifically any obsession to the opposite gender

Looking over your shoulder is a posterious lure

Jumping to conclusions stirs emotions in suspicion

Any origins to law culminates from paternal expectations through material domestications interdependence and provisional traditions in the law of the people

What is reality relative to what's going on that you don't know

Know any facts before drawing any conclusions

Facetious taunts seduce vulnerability

Violent and sexual content on television stimulates involuntary perverted thought a reprobate mind

Dreams never fruition without steadfast convictions

In life a sense of humor is essential' sets the comedic stage.

There's no such thing as a motion picture in time only your own minds eye recording

Inappropriate behaviors set precedence in depreciation

Thinking is the hardest task in life assembling any insight

Is the world watching you over your shoulder in expectation

Cognitive assumes cultural manageability

Time is the motion of change in perception

Notes:

Section Forty Four

There is no collective consciousness anyone knows

Social subjugation can be a blank stare

The undercurrent in life is paternal ambiguity

Reality assumes a collective association

Attraction lures chameleon gesture

No one knows any owns life recording in hindsight

An outlook on life is relative to a sense in sustainability affects the vibrancy

Family is relative to the ambiguous nature of paternalism

Why do some words not seem to be spelled wright

Reality tunnels drift into other lives stories stimulates imagination is most relative to domestic influences

Organizational skills are conditioned to any sense to sequence and order

Character is responsive to social subjectivity

Responsiveness is relative to any dissipation to the fog of expectations in any lures to seduction

Thinkingwhilereadingcanoverridecontextualcomprehension

Intuit pierces armor

Mixed company is an ambiguous association

Postulating control to any desired outcome challenges intuit

Normal is a posterious misconception in the variations to heritage

Artistic envy devotion

You tube validates common questions and influences pursuing answers

The forest is mystical in the wilderness sequesters orientation in exploration

The scent to life can sometimes have the aroma of artificial flowers

Role models influence appropriate social nuance

The air guitar is the easiest instrument to master

God is a mysterious association in the peripheral eye of reflection in silent radiation

Section Forty Five

Who in town has a bull's eye on you knows the same diffidence

Insecurity weights catapults of armor

Led Zeppelin is the greatest rock band of all time

Love is a desire in compliment and appreciation to who you know

Compliments lure expectations in identity characterization

Familiarity is a complacent factor to life taunts imagination

Nonverbal is the poetical undercurrent relative to motives in the layers to language in any attempt in communicability

No one knows anything no one knows

The precept to language is conscious relativity

Trepidation challenges social involvements disserting appreciation

New things in accessories is like a rebirth for the adamant consumer

Music lyrics stimulate unknown experiences in fantasy parroting roles

Dreams of being successful challenge well laid strategies

Stupid acts challenge retraction stigmatizing contemplative characterization

Dejection summons a fatuous bridle

Lust is a carnal obsession

Love is an intellectual magnet

What am I going to do for a living is a posterious potential

Tranquility wade's a stony brook

Jealousy yearns a poetic prose

Mischievous lures taunt scruples

Motive is the poetic undercurrent to seduction

Vulnerability is a utilitarian value

The internet is a collective conscious menagerie

Chameleon poses cast spells

Notes:

Section Forty Six

Value is any equilibrium to equality

Contribution has a sense in pride

Marital roles labor bounty

Being constructive has a sense of value

Responding to people who don't make sense makes one foolish engaging any bait

Empathy; you can never feel anyone's emotions but one's own

Motives are ambiguous speculations to contingency

There are no light particles only molecular illuminations

Addiction withdrawal is best met with fasting and exercise

A wandering mind is a dream in time

Cultural adaptation is a cause to contravention domesticating cogent repression

Authoritarianism is an auspicious paradigm

Much thought has a kaleidoscopic apparition sorting contextual associations to story

No one ever died saving people from the human condition to Cultural Adaptation

Materialism is an existential wardrobe

Sexual fixation is a carnal polarity

Authoritarians are ambiguous paternal mentors

Education impressions the illusion to a motion picture to time as a collective observation

Where is any social gathering after graduation

You cannot comprehend anything you do not know

If you know what you know about anything you will always be articulate

Buying things that last saves in the long run

Education conditions any bias national view

Insecure women cause the most antipathy in men contesting loyalty

Any malady is a paternal association

Notes:

Section Forty Seven

Metaphysics is the poetic undercurrent to our culture

Prayer is a posterious directory

Grudges are an obscure assertion dissipating in memory to the causes

Measuring success pursues any approval

When you look down on yourself you are looking up at someone wanting approval

Inalienable rights are a maladroit suggestion

Eastern eyes & vails seduce imagination

Authoritarian aggression is a maladroit protocol

When you are not comfortable with someone you are not in love

You don't owe the world anything

Childhood landscapes have mystical hues

Most dreams involve sanctuary and adventure

Experience in thought to language fruitions in pragmatic order like a quiz

Militaristic authoritarians know no defense

Conscience is a paternal warden

All things in education derive from philosophy

There was philosophy before science/physics

Listening to others stories strains thought in counsel

Assuming anyone you knew knows your life story is ostentatious

Mind is observation to perceptive relativity

Social subjectivity is an ambiguous barometer

The internet is a portal to collective conscious asscertations

You never know who you knew wanted to know you

Any merging insecurity has an open gate in fence

Credibility knows a certain consistency in action

Notes:

Section Forty Eight

Materialism is an opaque attachment to sustainability

Loyalty in marriage with children tempted by infidelity has an emotional sense to dedication

There is nothing to know; Consciousness is impressioned by our culture there is no preconditioned knowledge to mind

Loneliness has a cavernous inclination wanting the one in your dreams

It is so characteristic of people to blame current presidents for what previous admins left behind in the turbulent wakes of poor policy decisions

Domestication has a cultural disposition

Authoritarian nationalists are the most emblematic people who idolize patriarch monuments

Knowing how anything works involves applications

Forcing obedience is an authoritarian lashing in oppression to imagination and human spirit

Life is a mysterious fog until you realize there is nothing to know

What ever happened to the yard game leap frog or Simon says

What is intelligence? Law clauses a paternal sense to order

There is no intelligent design there was no knowledge before material domestication

Over stimulation is a grandiose escapade

Chameleon manipulation plays fools

Practice in vocabulary parodies experience

Ever want to memorize a dictionary

Maladroit seduction postulates a corruptible adolescent

Having structure keeps a mind in focus

Paternal antipathy as a parent is a yards yarn

Attention deficit can be caused by too much pressure

Authoritarian attitude is a posterious nose

Social perversion oscillates much involuntary thought

Subconscious is a cultural registry

Confucius says; one is an illusion in reflection in who one knew

Notes:

Section Forty Nine

In youth poetry is a daunting prose

Society is a web of confusion

Drunkenness is a limp puppet with bent strings

Authoritarian conjecture; paternal postulates intelligence

The main cause to domestic violence is insecurity

Awkward moments freeze frame

Social innuendo conspires characterizations

Possession is ego in a vice

No one knows what one knows only what one does gauging motives

Projected dreams instigate dispositions

Habits are conditioned by complacency

Humility is any retraction to ego

Much thought pages a derogatory directory

Inspiration knows no barrier

Much thought anchors prison bars to mind

Dejection causes cavernous antipathies

No one is going to make you anything you don't know

Expectation is a very consuming factor wanting to be a part of anything

Any world view is a media impression

Smart people question everything

Fickle is what fickle does

Character is an economic apparition

Dualism is the world looking back at you in expectation

Love is non emotional

Equal opportunity is an environmental disposition

Notes:

Section Fifty

Sexual innuendo in social settings is repugnant

Social subjectivity challenges demeanor

Any power in suggestion in media impressions to accepted behavior challenge sexual inclination

Authoritarianism is a chain gang master

Companionship desires are stealthy forces

Sexual desire conditions many emotions

Justice is the law of the people to maintain any sense to order

Transgressive altercations challenge memory remembering what happened

The force to law is transgression enforcing a civil code of conduct is a hazy nature

Reasoning is the quintessential diplomatic stage to pasture

Trivial memorabilia has any appreciation

You have to know a lot about people and culture to be a good comedian

You know you have hit rock bottom when you have been trespassed by the local trailer park

Any attachment has a form of identity

Social personification is any strange association

There is no such thing as normal is there

Love does not need any confirmation

Life is a glass menagerie full of obscure reflection

Love is a strange counter in attraction

Life is a struggle to being

Prosperity is the poverties child

Society has multiple cultures knows no bounds

Physical stimulation is an underlying cause to addiction

What one dreams and what one does is another thing

The crazy fool is in the library adding up the words

Notes:

Section Fifty One

Sanctuary is an untethered balloon

Life is a malady of pretention

Ego is an instantaneous exposure

When you are young the world is your own

Time is a gaff hook messes things up

Nothing is free you have to work for anything you dream

Things in common open doors to conversation

It's deceptive adjusting motive justifying any exploit

Life is a pretentious opportunity wanting to live in one's own dreams

Transgression is the catalyst to laws silent instruction

Remorse with the law has a shadowing aura

Youth has a pendant social aura

Pure thought flows like a stream of gentle water never ending from a natural pool of truth

The lack in trust is the basis to all failed relationships

The true world view is a social economic malaise

When life has lost its savor it's time to get out of town

We tend to see ourselves through the eyes of others or rather what we expect others to see in us

Reality assumes any truth

Change is always inevitable

Knowledge applies itself to the mind when anything makes sense edifies any misinterpretation

Never let the pressures of life wear you down always be yourself is a good suggestion

One would suppose being cooped up in a cave for twenty years would bring anyone to the same conclusions

Self-doubt lacks appropriation

A definition of ego is an ostentatious directory

There is an invisible bridge anyone can cross with an open mind to nowhere

Notes:

Section Fifty Two

While they wage war in a land so far away millions lay suffering in the streets of victory

Maybe there should be a twelve step program for news twenty four seven

Television media trains minds to five second sound bites' repetition confines focus

Rhythm brings the words to life

Wanting to be with someone special is a subterranean obsession

Existential is the assimilation to idiosyncratic fashion

Social behavior influences during puberty would be relative to sexual orientations

Imperfection has a seductive lure

Solving problems knows any score

Sharing music with friends has a poetic personation

Social rumors ionize idiosyncratic reflection

Having a talent is any good practice

It's not your world its mine

The real world is the one right outside your door

There is an inquisitive nature in adolescence asking the question what is there to know to be successful

Opportunity—optimism—venture

Behavioral disciplines are challenged by illicit seductions

The primacy to conscious objectivity is conditioned by cultural relativity

Exercise is the most important practice for better metabolism

Indigenous origins condition genetics

The analytical mind weighs outcome

A sense of being pursues meaning

Character is a posterious reflection

Resolves have a hollowing sense exhausting tensions

The Anarchist; Cultural resentment is relative to wanting the world to make sense

Notes:

Section Fifty Three

Trivial knowledge is a consuming factor

Imagination has many theatrical stages

Infatuation is a kaleidoscopic dream world

Telling anyone what to do contests competence

A sense in defeat has a sinking feeling

Conflict muster guards

Deception has a certain loss in value

Unbiased journalism in human transfer is a natural right

Manipulative lures charm deception

Consensual age independence franchises an authoritarian whip

Parental sanctuary expellation harnesses a material yoke

People who make things up have no memory recall to actual experience

Most things we think we never follow through on

Most memory is memorialization

To beat any addiction starve any craving

Education is the leading cause to social antipathy

There is no intuit to cursive slang

Creativity is challenged by originality

Anyone's mind is any impression to time

Sexual arousal causes a carnal obsession wanting emotional release

Love is intellectual knowing who you know

Money is the medium extension in forage propensity

Cultural prejudice is an ambiguous comparison

Social character is a maladroit chameleon Egor

Wanting to know any physics to technology is very intimidating especially in adolescence

Notes:

Section Fifty Four

Isolation is an abysmal cloud of confusion

Explicit sexual scenes in movies rouse carnal emotions

Cursive slang euphemizes derogatory exposure

Tally Ho!

Physical time does not exist time invented for making appointments

Identity has any retractable anchor

Perfection is a maladroit director

Blind ambition knows no course

Practice wants a private sector

Machiavellian people play off others ambiguities

In high school military occupation is the easy way out where there is no college prospect

Authoritarians have a false sense in security

There is no collective conscious mind there are however generational impressions in all native cultures

Social tolerance is a masked impersonation

There is no physical time time is an evolving culture

Fame has an insatiable desire

Character is a whimsical fashion in practice to any appreciation

Why does anyone require religious books when any knowledge is relative to human experience in culture

There is no intuitive construct to scripted memory

There's no true world view only what anyone follows like politics sports and arts

Thinking is a feeling or an expression of it

It is said that it is hard to get to know someone who doesn't reveal themselves to others and consequently people don't reveal themselves to you although it is very subtle and hard to recognize when engaged in our own ego

Denial has a defective eraser

Trust roles a level hand gaining respect from people

Desire is the cradle to all antipathy

Section Fifty Five

Competition is king and deception is everything

We all seem to share some wide view that binds us in a web of deception created through our collective view promoted through an illogical media depiction of reality

Insanity is being everywhere accept for where you are

If each person on this earth has a different view then it would make sense to say that there are as many countries' as there are people. Conformity to a world view has devastating effects on the human consciousness it is what creates conflict

It is the law itself that creates our crime. The more restricted our society becomes the more fear is generated and those wishing to escape end up at the gallows pole

Life is so much more interesting engaging people you meet provoking thought in each encounter

If Christianity is so righteous why does it always need defending? Nothing worthwhile need promotion, attraction is its best form of advertisement

Notes:

We need to do things for the right reasons not because there is some law keeping us from self-destruction but because we each see the world as our very own portrait

Rebel without a cause is one whom does not see the result of exploit in reverberation

"The anger in heart was created by the conflict in mind the duality of self "The Ego" The battle wages between two parts of conscious reflection, there must be two to have thought, did I see my conflict as the Mind processes vs. the body or flesh?

The body is the illusion the mind a conflict within itself intangible to the sight of body, the body or vessel to carry out the minds fantasy a god possessing the flesh.

I must look at the world and ask myself

   Does any of this, make sense?

      To what reality should I conform?

         And at what detriment am I willing to forfeit?"

Never let the pressure of life wear you down always be yourself is a good suggestion

A social view is only where mind's eye roams

Friendships are hard to come by cherish them well

Innocence can be lured by social seduction assuming posterity

Nationalism furnishes a false sense of pride

Any union is an artist's impersonation

Seduction is a peacock's parade

Crazy is breaking out of the mold is contrary

The queer eye is, who are you to anyone

Ego wants to be known

Possession knows a certain sense of insecurity where trust has no intuitive intimacy

Despotism is a malady of social economic disproportion

Paternal senses have a carnal emotion

Ego is any measure to success

Involving yourself in others relationships is an kaleidoscopic medley

Notes:

Section Fifty Six

Why do some people require defending certain behaviors

Memory is an isotopic apparition

I'm sorry knows no intuitive conjecture

Democratic cause is an economic conundrum in any strife for equality

Capitalism is the number one cause to social dissoluteness

If you are working under the table in construction you are shooting yourself in the foot for unemployment benefits and workman's comp insurance

Cynicism postulates a common knowledge

Ego is the world looking back at you wanting approval

Taming masculinity, insecurity, confidence, intimacy, dominion, supremacy,

Negative people radiate a nefarious persona

The consumed mind limits focus and comprehension in all people

If you know what you knew in retort no profile will loiter

Authoritarian justice is a retaliatory lashing

Adolescence resents alarm clocks in adaptation to compliant regimentation

Taste stimulates sensual perception is the leading cause to addictive substances

Cravings cause obsessive compulsions

Economic partition is a mystic warden

Pessimism is a shackled disposition

Memory' no one knows what you knew

War on words is a literate blunder

Dream worlds are anyone's own illusion of time

Any collective aura is any circumstantial phase

Virtue; you will always know what you are doing pursuing intuition

Progress would be maintaining a common wealth of interest

Peering stature is an isotopic strain

Notes:

Section Fifty Seven

Circumstance conditions personality

Education; no one can master two arts at the same time unless they merger

You want to have a good relationship with yourself before having a good relationship with anyone

A world view is one's own conformed polarity

Carnal compulsions are the main source to societal perversions

God is not the author of confusion, god is any intuitive juncture

Knowledge edifies experience initiates any sense to resolves settles mind in reflective mediation

Economic turbulence ensues societal insurrections stimulates global conflicts

Social objectivity is a Pandora box

Economy trains mind to expectations

Authoritarian law imposters intuit

There is no collective mind only any selective impersonations

Solitude is when you close your eyes and separate yourself from all the noise

Childhood is a mystic orb of confusion

Loyalty knows certain virtuosity

Involuntary thought is an estranging apparition

Freedom is a state of mind in animated suspension

Time is a temporal condition in provisional directories

There is any frustration where others simply do not follow the same practices in household edicate

Thinking is any art to insanity

Intelligence knows what we are as a culture

Being genuine knows a sense of dissonance

Innocence surveys any merit

Feminist campaigns provision value to contribution

Jumping to conclusions muses traits

Notes:

Section Fifty Eight

Homosexuality is a carnal parody

Popularity in school is an ambiguous posterity

War makes people stupid

Domestic abuse in childhood personifies social antipathy

Childhood influences challenge authority

Memorialized views consume minds

People pleasing caters discordant behavior

Exercise reduces stress abilifies coordination

Authoritarian aura libel's culpability

Happenstance lingers acquaintance rewind

Love is a harmonious sequester

Prejudice is an ambiguous director

Life is an eclectic orchestra

Wolf in sheep's clothing is a fiddler's swindle

The real world is a pendant harp

Infatuation is a dreamer's lure

A walk in the wilderness is a mystic journey

Human physical mediums include caricature to fish and birds

Insolence knows no diplomacy

There is no natural selection in human procreation human procreation is relative to genetic transcendence wanting a part of you to regenerate

The big bang was a cosmic procreation

Property; losing anything to theft is disturbing

The mystery to life is a mystery

Success is not an achievement of a goal but the perseverance of action

Ego wants to show you something

Obesity; eat vegetable's only and anything bland to lose weight and walk every day to help metabolism

Notes:

Section Fifty Nine

Negative media barrages desensitize human compassion

Before the printing press 1430 there was no social media keeping track of the world, there is no true accounting only any original documents all else speculation

Royalty is auspicious in any postulations to authorities of intuitive virtue

History lessons should not be graded in school assuming there is such a thing as a true world view

People think themselves stupid thinking there is something to know

The world view is a challenge in perception to innuendo

Discrimination postulates any normal

Famous people are not always the best source for intuitive assessments

Who you know with anyone familiar is one's own dispositional polarity

There is no outer view movie only your own selective interests

Humility knows any cause to sound reasoning

Government official postulates democratic decisions have common interests

Social class pretension presents an aire of envy

Courage is the virtuosity of indifference

Economic environmental depression is an oppressive atmosphere conditions posture

There is no collective aura

The law is a maladroit warden

Solving problems is a patient endeavor

Economic position conditions bias perspective

Existential; property is any physical medium extension has any emotional attachment

Self-value has a sense of appearance in an image driven society

Laziness has no reward to achievement

Social involvement is a crap shoot venture

Personal value is a paternal domesticating proposition

Everyone has their own secret world

Notes:

Section Sixty

National global media conditions social focus

Paternal dejection postulates exclusion

Sexual innuendo impressions strange thought

No one can comprehend your experience is an estranging consult

A temporal congestion is the cause to compulsions to self-medicate

Intuit knows common vulnerabilities

Desire leaves a left temporal impression

Material adaptation is a maladroit yoke

Acting has a posterious retraction

Who or what assigns social responsibility

Socialism is a libertarian cause weighs balance

When you do things that feel right you have no remorse

Drama is a manipulative seduction

Nothing is what it is until it is what you know

Exclusion is one's own illusion to a social consideration

Liberty is an uncorrupted mind

Responsibility is a double edge sword conditions time

A world in motion is a posterious involvement

When you are listening you are thinking construing any dialog

Knowing how to know has much to do knowing what goes on around you

Disappointment is an estranged dream state

Association prompts opportunity in occupational genre

Women can be in public view in a bikini; however a man would be arrested wearing only his under pants and we thought it was a man's world or is it

Lusting for youth age is a state of mind

Premature relationships limit independent potentials

Notes:

Section Sixty One

There is no such thing as a collective reality tunnel recording

Weather conditions ambition

A car is a human medium extension of mobility

No one knows one's own characters projected lure

Being successful postulates the world's approval

Ever wonder what you will know in the future

Reality would assume something we all know' there is; "confusion"

You cannot lose anything you do not have

People who assume control of others are despotic vipers

Courage endures calculated risks

Material procurement conditions most thought and projection

A changing collective social mode is an economic barometer

Most relations are superficial conditions intuitive stratagems

Much experience practices no realization in self-dialog to existential knowledge

You have no control over anything you don't know in people

Original challenges the heritage to art

The lure and pull of the world conditions focus limiting the art to any quarantined practice

A government hand is an erroneous custodian

Assumed authority in adolescence conditions corporal polarity in expectation

A collective veracity is a nation's global relations in bias positions

Electronic leashes distract focus

After high school with no future education you realize you have no job experience

Social pretention assumes any maladroit stature in comparison

Lusty endeavor is a fool's bridle

Imperfection is a seductive fashion

Notes:

Section Sixty Two

Authoritarian is a strange mentor assuming guarantee

How you know anything conditions distinctiveness

Most cursive slang is stimulated by incompatible stances

Linguistics ares the languages to each means functions

Quantum theory is relative to human existence relative to cosmology

Freedom to speech exists being there is no interpretative cause

Media technologies stimulating subjectivity challenge comprehension knowing cause

Individual reality tunnels are conditioned to the perpetuity of a changing sophistication

Conversation lacking dialog is congestive

Identity has many forms in self-reflection

Many words have no physical connotation where definition relies upon memorization

All practice has a sense in failure

Thou shalt not think

Thou shalt not ask questions

Thou shalt not question authority

Thou shalt not assumes a master

Most conversation has no intuitive construct

Get out of my head and get out some Led

The real world can be very hostile

Television addiction limits imagination conditions time

High school graduation dissipates a sense of union

Trust knows a certain consistency

The only thing certain in life is your own expiration has a consuming factor

The force in what we do having no intuitive guidance conditions temporal confusion

There is too much emphasis to having wealth as being successful

Notes:

Section Sixty Three

Class reunions have a sense in redemption

You cannot know what you remember in others experiences

Patience knows no time

Stroke the ego' Egor

Karate Kid; Transgression is sinister do not lose to fear senses shame for not being stupid

The American Civil war was stupid…

Democracy knows no cause what is "is" economy/free trade capitalism where control is any regulation

In motives not knowing what's going on around you tracks no involvement

Governing representation privations definition to reason

Politics is any controversy to resolve to local national and global determinations

Empathy is a paternal lure

Complacency lacks incentive to reward

The world view is a media impression reality is your own domain

Frustration anticipates the fruition to result

Wanting to impress people lures a spot light

A friendly voice has any attraction

Authoritarian cogence is counter intuitive in premise to paternal origin

Love is equality in appreciation

Patriarchal senses have anything to do with looking over the shoulder postulating contribution

The social club has a luring pull challenges association

Human predictability suggests common influences

Most people simply become programmed to what they do and then there is any appreciation to any art

There is what you do and what you look forward to

Known deceptions provoke confrontation

There is no such thing as what you know as a collective state of mind

Notes:

Section Sixty Four

Conformity is authoritarian submission don't ask any questions rocking the boat wearing blinders

Luring recognition is an egocentric lasso

Reality is any connection to common occurrences

Consumed thought is conditioned by life's economic pressures

You can never know another's perception of the world knowing anything conclusive to cause

There is no collective world view recording time there are only idiosyncratic perspectives

Focus is a station in time

Desiring approval lures ego in appreciation

Alcoholism is not genetic it is a mental and corporal compulsion conditioned by a congested state of mind

Not knowing what you are part of ostracizes intellect asking the question; what is there to know postulating authorities

---

Nouns have no intuitive construct are simply memorizations to idiosyncratic identities

People are expected to respond a certain way to national monuments assuming an idiosyncratic persona in patriarch bias to postulated integrities

Drunkenness has a degrading sense of being

National pride is an insidious pitch feigning achievement

Intuitive construal fruitions any cultural relativity

Corporate consolidations cause eruptions in displacement to people's lives in loss of jobs

People will try to make you look stupid when you ask questions about the culture

Exposing your vulnerabilities in a relationship while young sets precedence to codependence

Most thought concerns money

No god know god

Advertisement is very alluring

Mind is any inspiring mode

Motivation is inspired by interest

Admitting to yourself you did something stupid is healthy for the mind

Things are often opposite to what they seem where envy to competence may be reflective

Don't think' let government do it for you

Notes:

Section Sixty Five

A conformed view is a conditional sanctuary asking no questions

You cannot love what you do not know

Authoritarian is a regimented drone to obedience

Nationalism seduces a sense of honor in respect to illusory causes

Any compulsion in addiction is a very degrading experience in lack of discipline

Astuteness necessitates any leeway

Children sense parental intimacy is a healthy home sets good examples

Moral infers a social standard; what are common senses'

We admire people who make sense

The world view is a fuzzy screen with an intangible dial

Opulence is the nature to knowledge in who you know

People are governed by pleasure and pain

There is no all seeing eye there is no collective conscious mind

A social mind is your own changing perspectives asking the question; what was that

Any dialog is best posed as any question rather than a lecture otherwise it is difficult to know what anyone follows without context providing clarity to any tangibles

People become addicted to the news anticipating any resolves to the worlds never ending conflicts

Buyer's remorse it wasn't what you imagined when you realized it was a compulsion wanting something new became old the day after

If you sense any dominance or are being dominate you are not in a compatible relationship

You cannot vaporize living organism and teleport them though a machine to reassemble them in another location contrary to what physicists theorize in teleportation

Memorizing functions is a good practice in comprehension

Infatuation; I didn't really love you I did not know who I knew only who I wanted

Surveillance society is the perversion to capitalisms suspicion

Church provides community gathering

Jealousy is the seed to social antipathies

There is fantasy and then there is the real world of confusion wanting to make any connection

You want to judge people by their character

Notes:

Section Sixty Six

Television is a maladroit accessory

You get your morals and values from your own common senses not religious belief

You do not know anything you have been told

The cause to philosophy is what reality is, reality is what the culture does

What is it with trust with people?

Timber industries are an awkward barber

Law does not change the causes to society perversions

Statistics; there are likely as many unemployed not being accounted for who were or have never been approved for benefits

When you cannot remember something you know your mind gets stuck in a fog

You cannot change people you offend them when you try

There are only the people you know and you do not know their reality you only know what you experienced together with anyone though there are still two versions

Naivety and trust are synonymous

The price of gas changes the mode conditions expendable income

A social mind assumes a conscious connection what is there to know?

A police state is when anyone is under suspicion for questioning authority what is it

Regarding requirements money conditions a perpetual strain

Regarding law losing your driver's license is a very degrading experience sets things in motion is an occupational hazard

Letting anything go is the path to a free mind

You only know what you know not what is assumed

Any stimulation conditions perception: taste is the number one addiction

The means is a very degrading experience for anyone especially anyone living through impoverished heredity

When you don't know' what you said there's a retraction to thought

Most television mimics perversion

Inflation: Minimum wage wants to be consistent to cost to living earnings ratio

The corporate stamp conditions formatted views

Notes:

Section Sixty Seven

Trying to describe reality is like trying to pinch specks of dust floating in air

Any stern façade regards something is askew

It is always healthy to find the good to any times

Like prayer chanting does nothing for the mind

Ever felt embarrassed for anyone

If a relationship does not feel right pull get out now

When anything you were involved in didn't make sense the mind plays it over and over

Being drawn into conflict challenges audacity to be indifferent

You do not want to be a ward to the judicial halls revolving door

If something makes sense you can trace it backwards finding it there again

Failure is another word for practice

Is sex embarrassing for most people, movies make sexual performance seem casual

I don't know myself to well however I am my own conversation in mind

You want to diffuse any situation pacifying emotions to any antipathy

Economic negativity tightens purse strings

Anything that does not make sense likely isn't true what is does

Regarding memorization, grading is degrading and classifies anyone

The arts describe the world view is a cultural epitome

Liberty is being your own person not another's possession

Liberty regards dominion sanctity is your own person

Glory mystifies the view

Regarding people memory is essential knowing what you do being consistence

Who knows who or how you know anyone assumes any association

Media views imposter a collective mind

Common distresses in confirmation is emancipatory

Notes:

Section Sixty Eight

Thinking anything through before moving forward is constructive

Measuring anything impressions dual persona

Expectation comes from living up to any heredity to traditions

When you get high you want to apply yourself to something creative

Having a criminal record stigmatizes the view there is no cultural emancipation changing the causes

Philosophy classes having open forums practices social rehearsal

Constructive has any application to talent wanting to be successful

Regarding the weather people always come together in times of crisis

What was I just thinking that did not make sense to me' ever had that thought

Self-knowledge has a positive affirmation

Aggression and insecurity is synonymous

No one knows how anyone knows other people assuming a true world view

Regarding drugs withdrawal initializes the healing process causing fatigue

The Beatles were the most influential band in expression to a cultural transition

Driven people are consumed wanting to be successful

There is no physical conscious connection

Time is any anchor to historic inference

Time is the expectation to goal the journey is the goal

The human condition is inhibitions having no instructions to cultural adaptation

The cultural urban view is Van Gough has any luring charm

Inflation dilutes appreciative value

Advertising jingles charm loyalties

Franchise is corporate mafia capitalism is a surreptitious freedom

Self-deception is a Pinocchio's nose

Trust is an ardent warden

The lack of funds initiates many degrading thoughts

Notes:

Section Sixty Nine

When nothing makes sense there is nothing to know

You have two lives what you are and what you are part of

There is no way to describe a future as movies impression what is is

How does one fit into a culture where there is no definition

Youth is an invincible dare devil challenging mortality in rebellion

Strength and endurance is a social challenge editing the drama

Equality is a fair shake

Ridicule infers a posterious imperfection

One cannot change culture one can only interpret things as they are

What you do affects modes

Egocentricity garnishes approval

Normal has a material conviction

Euphoria is a state of mind in grandiosity being the essential cause to addictive tendencies wanting to be in a state of emancipation

Regarding nutrition who knows how anything is healthy

Honesty knows vulnerability

In youth having no structure in dysfunctional settings dispositions institutionalizations

Existentialism; everything you own has an emotional affiliation

Consciousness has no objective yet you know you are alive

You only know what you experience

Life is a struggle for most people what does government do conditioned to policy

The moment is often in projection to a future in allusive dreams

A pragmatic collective mind is an economic gauge

Authoritarian assumes a psychological warrant constitutes malpractice

When life has become too plastic it is time to light a candle and melt away

Damn Weather

Prejudice affects stereotype to idiosyncratic oppressions

Notes:

Section Seventy

Thinking yourself a fool assumes anyone knows what they're doing

Everything is an expression of character

A little house on the prairie was a lonely oasis

Vocabulary is a cultural menagerie

Perfection wants everything in order

A monetary bondage for a house wife care giver wants a paternal barter

Times of crisis muster defeat or perseverance

Without family and friends what is there

Thought overlaps sentences assembling what you know

Getting high (grandiose euphoria) for the first time has a sense of emancipation becomes the obsession to escape a repressing aura of expectation in the cogence to adaptation

Fatuous people manipulate others based on knowing the same inhibitions

Conscience is the observer of thought impressioned by cultural influence

Physical mediums characterize Id

Speculation has wide ranges wanting to know anything

Desire drives ambition

Expectation conditions time

Any perversion in thought has a second glance

Shitt's Creek was a stupid title for a television series

Measuring anything impressions dual persona

There is no place to be on drugs there is only knowledge to cultural impressions

No one knows what anyone knows there is no true world view

No one knows any others life perspective

What you say and what you think isn't always the same thing

There is no collective mind of knowledge invents gods associates to existential dualism in the world looking back at us

There is no equality

Environments condition influence

Notes:

Section Seventy One

Wealth & fame distinguish no virtue

No one knows your world no knows who you knew

There is no comprehension to others experiences listening to stories

Intuitive art tells any real story

Thinking anything through before moving forward is constructive

Sequestered time develops any talent is a challenge in dedication

All people make a certain impression

Cultural linguistics what are they

Anxiety lies in the diagnosis of the prescription

Doing your best is your own distinction

Stir fry vegetables have more flavor best served cold

A person's value is characterized by what they do

Praising a child caters to style in approval

Letting anything go doesn't mean anything unless you've reformed your own expectations

National pledges charlatan any alliance

Cyber space; we have reached the climax to information overload

Nothing grows on a beaten path

Wanting to have someone too well is a stirring emotion

Believe, believe, and believe anything you don't know

Not wanting to use a condom defies a certain masculine consummation

Get Smart' Control to Chaos

Familiarity stimulates desires for new and improved

Poor economies increase violence often times the wrong ones end up in jail

Using the thesaurus is a good practice in comprehension

The age of innocence endures a resistance

Why does concubine come first to mind wanting to be with someone

Section Seventy Two

Companionship has a confidant trust

Any natural right is to be one's own mind

Bringing up others children as your own is a noble virtue

Loyalty has certain karma

Aggression and insecurity are synonymous

Crafty people are always changing the story

When a person writes one assumes to share dialog with a collective mind

The social sense is the mind in appraisal and comparison

There is no unified global supervision regulating environmental causes to pollution

If government is anyone's ally why does anyone fear and loathe it

Small towns are big when you are small

Fame places expectations on people to be role models to virtues

A reputation is one's own personal interest applied to conduct

Telephone rings prompt corporal leashes

Smoking and other habits condition thought to interval breaks

Taking charge of any project applies cognitive organizations

Searching for lost items conditions conviction like a grouse hunting for seed in the same circle

You can only be with yourself no one knows who you know

Spousal domestication has an isolating factor in social limitations

Desperation is when every inch takes a mile

There can be no way to describe the mystery of life instigates mysticism

Suspicion scrutinizes sincerity

Life sucks when nothing makes sense

There is too much pressure to survive for most people

Nationalist institutes entice unionized character endorsements

Sincerity has no remorse

The Wizard of Oz... The Lion represents "Courage," The Scarecrow "Heart" and the Tin Man "Mind..." conditions Virtue being one mind is the theme to the New Testament

Notes:

Section Seventy Three

Having a positive influence earns respect

Sodium and salt effect constant stress

Indigenous people have more unified features

Wall Street initiates an economic pulse is the artificial life of the machine

Cigarettes are a low dose narcotic why are they legal

Life is all about when you can count on a well prepared meal

Writing is the best way to practice intuitive dialog

Jealousy is often sensed backwards postulating supremacy

If you can't get over a fence go around

It's difficult to recall anything accurately contributing to any disturbing experience

Legal shysters Duey Cheatem & Howe want your money

Judicial buildings have a certain aura

Vengeance has a chameleon implication

Embarrassing moment's freeze frame

Keeping a diary in reflection edifies life's confusions

Red alert read alert Stranger danger Stranger danger

Movies impression all the parts know a same story

Salt is the most addictive substance

Material culture has no meaning though has a sense of sophistication

Law is always right there watching is a posterious shadow

Celebrity impressions the view envying posterity wanting to be popular

It would be nice to stick with the things that work rather than always trying to reinvent the cart

Allegiances to flags is an ambiguous doctrine

Everyone is a stranger until you get to know people

Memories are kaleidoscopic peepholes

Describing anything practices intuitive comprehension not mathematical equation

14 is the age asking the question "what is psychology"

Notes:

Section Seventy Four

Anything you eat having a bite or burn may not be healthy or nutritious

Sometimes the obvious is right there before you being rejected oscillating context

Having any talent requires dedication and practice

Beavers are clever creatures

In general people are not stupid they just haven't had the best education

Honesty cleanses thought

Ultimatums are manipulative forums

Diplomacy knows any common denominators

Anger' something did not work out the way you wanted

The collective view is an economic pulse always in recovery

People having no money is a constant stressor

Unemployed middle aged recons much discrimination

High School counseling is essential knowing options

Popularity is a strange expectation

Assuming authority prompts any expectationWe all know the same thing and do not know how to pronounce it; "SEGREGATION"

Arming teachers with guns infers there's something awry in education

A trivia query; what played cousin "It" or "Thing" in The Adams family

Personal domain has a toll gate

Advertising sympathy ensnares empathetic causes

Ego has a mysterious coercion

Over indulgence is an inability to be conservative in anything

Trying to make sense of anything having no resolve anchors mind in rhetoric

We are convinced by education there is a collective mind

Progress is a tethered towline into an abyss

There is no time there is only one physical existence

Notes:

Section Seventy Five

Drama is a maladroit paternal convention

What you do to yourself on drugs is nothing but stupid isn't it

Comedian's council inhibitions

Where there is love there is devotion

Dogma instructs memory

Statistics; an average household annually sends $1600.00 of perishable foods to landfills?

Addiction loves company

Charity infers an economic obligation

Bland foods are less addictive

Have you ever gone along with things making you uncomfortable

Childhood trauma has any penitent fog

Marriage imposes a subsisting trunk in wardrobes wearing any role

Lower alcohol limits boost municipal revenues

Drinking ample amounts of pure water aides immunity reduces stress

Children make up stories to feel apart socially to draw any attention

You have a paradigm shift when you leave your home town has a change to identity like a new start

Fantasy rouses any rehearsal

A social mind is ones own view

Same sex marriage is an awkward correlation

Gullibility poses anything is known

Manipulation knows no intellectual interlude

Pleasure in life is the art of being what you want to do

Nationalism imposes memorialized character distinctions

Poverty solicits a social idiosyncratic apparition

Crime shows impression suspicious characters

Time is an age in evolution

Notes:

Section Seventy Six

In multiplicity to cultures no one has no clue what anyone knows

Youth has a certain swagger

Eating breakfast induces lethargy affecting concentration

Dancing stoned is the spice to life

Greed influences inflation intensifying strife in civilizing stature

Social antipathy is the number one reason students take guns to school

Television conditions much perverted thought

Falling in love in high school having no income is difficult stirs ardent desires

Character shades a lingering posture

Capitalism is a class menagerie to posterity

Innuendo pinches the gourd

You are your owns health care provider knowing how anything affects you physiologically

Law assumes a certain disposition

Is it fear which drives ambition or is it gratification

Anticipation is pressure for any conclusion

If you do not read or write you will not expand vocabulary

Commercials with women eating pulled pork is not sexy

What you do constructive has any appreciation

Chaos is the cost of symbolic freedom

Trust infers certain vulnerability

The first time you drink and lose your faculties causes a stigma

Respect trusts your own intuitive disposition being one mind of your own integrity

Television crime shows impression the law is sincere' is it? Well that depends on who you are'

Organizational skills manage efficiency knowing what you are doing

Life is a consuming treadmill when you do not enjoy what you do

Notes:

Section Seventy Seven

Appreciation concerns any copacetic moment

Jealousy has many venues in character expectation

Television allures envious lifestyles

Having a good job affects aptitude

Movie vulgarity is not sophistication

Facetious behaviors subject any challenge to demeanor

Fatuous people want to make you look stupid to yourself

Contagious illicit behavior is a fool's haven

Resilience has a pragmatic conviction

Time is the obligation to a set parameter

Movies impression unrealistic life styles

Perfectionism; one is ones worse critic

Being too generous is a posterious invitation

Modern conveniences obsolesce old ways to inherent skill

Making close friends is rarely often

The world expects you to fall in line and pull your own weight' is us in disguise'

Concerning social lures image personifications glamourize idiosyncratic fashions

Existential interdependent dualism summons an authoritarian apparition postulating knowledge

Indigenous temperate agricultural origins condition cultural occurrence

You Rang…. Lurch is on the lose

Foods causing a lethargic reaction may not be nutritious

War has a strange sense of honor

Authoritative devices provoke emancipatory causes

Fasting relaxes the digestive tract starves out addictive cravings

Depression is a state of mind where there is no sense to adventure

Notes:

Section Seventy Eight

Unsolicited advice checks competence

Memorialists' lures pose any anchor in benevolence

Parental dependence has a safe haven when you don't burn any bridges

When you do anything stupid you want to block it out of the mind

A gun is a physical medium extension in aggression

Confucius say' you don't know' anything you don't know

When you have nothing to look forward to any life becomes dreary

Law postures rear view mirrors

Law cultures boundaries

Facetious conducts taunt injurious boundaries

Sodium really does reduce endurance

Reality is only what makes sense to you

Raising children is very consuming demanding patience

Poverty has two parts social and economic strife

Laws do not get to the causes of addiction

Involuntary thought in sexual obscurity to equivalent gender are culturally suggestive orientations

No one can read your mind no one knows your reality

Humor is essential if you want to have any peace of mind

Fear of being successful practices failure

No one knows another's perception of a world age

Vengeance has a maladroit convention

Tragedies pull people together

Thought has a broad spectrum

Sodium is a leading cause to addiction

Kung Fu guards no fence

Notes:

Section Seventy Nine

Arrogance is a posterious authority

The mind wanting to be a part to anything is emotional suiting any role

Family is the social gathering in one's own clan

If you grow up without parental role models you tend to be more introverted

Postulating consensual age law charlatans sensibility

Idolization stifles personal ingenuity

So many films impression that sex is the ultimate goal

Regarding youth naivety and innocence are synonymous

Attraction has an emotional compulsion

Running low on money has a sinking feeling

Posterity is a mysterious echo

There isn't anything you do doesn't costs money

Every City has its own idiosyncratic presence

Social trepidation oscillates democratic yearnings

Television initiates many views influencing thought

Concerning history films invent conjured scenarios

Children like cordial surroundings

How certain people treat you poorly can be somewhat a mystery assigning motives seducing martyred reflections

There is no communication where there is no contextual dialog

Government is a maladroit suspicion

Philosophy distinguishes cultural nuances

Attraction has a denying resistance

Expectation concerns an awkward approval

Claiming statistics most theft goes unreported

Law is a maladroit sibling

Cultural equality conditions any balance to liberty

Notes:

Section Eighty

Pessimism weights an anchor

Personal secrets and embarrassment are synonymous

First acquaintances are awkward until you know anyone

Cop shows are a social perversion

A social malady consumes thought limits focus in education

Life is in a barrel tumbling down Niagara Falls a daring adventure

A camera is a physical medium in perception to visual memory capturing the moment

Modesty has reward

Good acting improvises any linguistic dialog

No risk no reward

You have the right to believe anything you have been told

The fool is one who acts before thinking anything through

Complements measure character

It is the art that we choose which fulfills our dreams

Fear of failing has a burning sensation

Expectation congests thought

History lessons practice idiosyncratic comprehension

You remember what made sense all else just fog

Reality is within any knowledge to what we do

Using a thesaurus distinguishes definitions

Elucidation is when a light comes on in your head when anything registers

Pessimistic tones lure challenge

Life is a dream in random composure

You use construal diagnosing car trouble

Religion is all speculation

Notes:

Section Eighty One

Combat tanks are suits to armor conjoining infrared graphics desensitizes war

Video games are mindless endeavors

Throwing in the towel is when you stop caring for yourself

If you care for birds why place them in cages

Memory is mostly cultural robotics

Honesty has a certain tone

A suited companion nature's character

Opinions instigate definitions

Outlook renders any canvas

When there is any crisis people pull together presented opportunity for charity

No one knows what anyone knows as a true world view only prominent fidelity to "global emergent" ideological divergences

Economic prejudice garnishes market share

You don't know how you are known by anyone you are never who you knew

Speaking in groups is intimidating unless you know what you know on any topic

The local news presents social and economic currents

Government presumes a collective goal

You're not insane there's nothing to know the world is fatuous'

Money conditions everything' bound to it you are a slave

There is nothing to know' consciousness is conditioned by culture

Social transgression makes people stupid

Being on any prowl has a mystic notion anyone wants to know you

There is no government guarantying anyone an occupation but politician's

It's no mystery why people murder others it is elicited by social perversion and rage

Having any mechanical aptitude in youth is a great asset

Education impressions a collective mind everyone knows is "monumental" bias significance

Notes:

Section Eighty Two

If you believe in an afterlife any headstone should read

"TO BE

CONTINUED"

Authoritarian feigns a responsibility honoring no constitution

Psychology is environment conditioned by traditional models not corporal designs

Judiciary is an inelegant virtuosity' material culture is non instinctual it is trained

Gangs are poverties patriarch institutions

God and law assuming a divine authority is not science law imposes a moral conviction is religion, there is no separation between church and state is resolved by science

Education poses national idiosyncratic nuance postulating historic inference

You know you are sane when you can't stand the friction people around you swirl

Asking questions during lectures is important "grasping any context" helps others probing the same uncertainty

It is real and it is sometimes raw what people around you go through in the urban jungle

In youth invincibility plays a mysterious contest associated to memorialized agents

You want to tell your life story expressing who you knew

In new acquaintances if anything doesn't feel copacetic trust your senses

The splendors to traditions in heredity are any social gatherings

Anniversaries force spontaneity in expectation postulating reward

Collection agency phone calls make us feel like dupes

Retirement in a camping trailer is not the American Dream is a blue collar cage

Gold fever is a compulsive gaze

In one way or another anyone can feel like a failure

Appeasing anyone is a codependent snare

Mean what you say say what you mean

Films impression stereotypic ethnicity; there is no national character anywhere

Idiosyncratic gestures chameleon impressions

Movies conjure imaginary life styles

Inflation is a derogatory nuance

Notes:

Section Eighty Three

Drinking beer affects the sphincter causing stress conditions the bladder and may contribute to prostate cancer

Attempting to change people you make things worse posing supremacy

Transgression postulates incompetence assuming authority to intelligent design

Letting anything go is the path to any free mind

Like prayer chanting does nothing for the mind understanding does

It is always healthy to find the good to any times

Are you owning or being owned

Shock value is classless exercising freedom to seditious expression

Hot peppers are carcinogens

Intrinsic art challenges inhibitions

Raising kids is inhibiting posturing what anyone experiences

Greed and poverty devastate lives and an innocence of all children

Speaking in any public forum requires practice

Life can be a pressure cooker

Women celebrities fashion is very sexually suggestive what's up

"Feels absolutely disgusting" Anorexia is glamour's child

Ladies fashion extravaganza is extremely audacious

The world is raw especially if you are young and single

Greed is a mysterious manner

Men cry when no one's around

Pledging any allegiance to symbols is mindless idolization

If you suffer from insomnia you are over stimulated

Forage propentence is instinctual

Grade school is a frightening experience involving social antipathy

The law does not know who you are to people who know' you are any jury of your own peers

Section Eighty Four

The camera was the most expressive invention

Colorization to film was a big deal

Pretension is a posterious shadow

Ego's posture any challenge

Much sixties and seventies rock music is very mystical

Finding the words for describing certain things can be painstaking

Dreading anything consumes thought

Identity signifies character roles

Inflation dilutes quality

Expectation often contributes to double standards

Sibling rivalry posturing favoritism is ambiguous

Pessimism is a plea for comfort

Compliments lure a posterious perfection held by attraction in paternal ruse

What are wave functions correlated to

Age twelve the world is mysterious concerning what to know

"You" are your own property nothing owns you

Happenstance has a rear view mirror missing a connection

Any addiction to stimulation is depreciative

Any domestic malady is antipathy knowing what's going on

Adulterated rage knows a cultural displacement to forage subsistence

There is no one mode social mind, media imposes any current sentiment

Where there is no adventure there is no reward

How would you feel if someone you knew was addicted to heroin?

Controlling people is a sign of insecurity

Food is a very sensual experience

Notes:

Section Eighty Five

Technologies merge cell phone camera internet television "all in one" device, what's next media implants

Complacent stress causes over eating over eating causes stress is a vicious cycle

Letting someone go as a job is an arduous task

Reciting law has no intuitive virtue

All people have common inhibitions

Honesty has many digestive considerations

News Media's condition anticipatory psychoanalysis

Regarding law mis-demeanor infers something'

In conversation connecting any dialog knowing what to say to anyone can be flabbergasting

Hair style is a consuming consult

Social vogue is an inelegant trophy

Video games condition motor memory aggression

Hollywood impressions time as idiosyncratic fashions in vogue stereotype

Bleeding is very frightening for children unaware of the body's own healing powers

"Blood platelets" an itch is when something is healing

Being punctual is being conscientious of other people's time

Vocabulary is the physics to cultural epitome English is the most comprehensive language

Cultural linguistic barriers mis diagnose ideological controversies

Having anything to look forward to aligns focus

Law assumes a common competence

For many' television is the only social medium

"I wish I had that to do over" ever had the thought

Being wrong about anything evaluates motive in appraisal

A social world view is anyone's own perception

Disappointment is an ambiguous engagement

Notes:

Section Eighty Six

Certain foods rouse chemical stirs

Fasting; abstinence from all foods and chemicals rests the whole body impeding repetitive memory

Practiced mechanical skills instill pliable confidence

Children mimic craft

In envy what you think may be mirrored in reflection

Children appreciate being spoken to minus condescendence

One mind is what you knew when you were free as a child being provided for

There is no way one can reason with arrogant people

In general we all know the same thing we don't know a common cause

You feel stupid when you can't remember something you just knew

There is no physical time

Movies impression sophisticated dialogs that do not exist in normal life

Per capita poverty has a higher rate to teen and unwed pregnancy and addiction

Trivial knowledge in cultural memorization has pretentious shades to dialog

Poverty and addictions trigger combing vultures' armor defensive virtues

 "Logic" was the main theme in the seventies

There is no defining moment to a unilateral ambition

Democracy is a maladroit convention

Any social stage can be inhibiting

Main stream media caters celebrity envy

Television drama has a desensitizing affect

When you say what you mean there will be no lingering shadow

Currents to arts progression impostures collective ambience postulating one mind

If you know something it will always come back to you when prompted

Alliances are estranging postulating association

Notes:

Section Eighty Seven

You always want something new unless you have your own unique style

Curiosity stimulates exploration

A good plan has any flexibility

Natural law is the instinct to survival paternal's any species

A little voice in your head telling you how to think is ego conscience

A driver's license makes you a ward of state law is an awkward privilege owned by a thing

Better memory is unreserved always knowing what you do

Sometimes the best things in life are right there in front of you always there to be appreciated

The Ocean is not just salt water it is largely pulverized shale which has accumulated for millenniums including ground crustaceans

Much education applies no ongoing practical application

A confident presence conjures posterious tolls

Municipalities are always going broke reasons why police are always on the prowl

Every economic position has any owns prejudice

Celebrations blossom any ambient mode

Cultural geography impressions topographic memory

Education is a specialized art to any applicable appreciation

Is sexual performance embarrassing for most people

No one knows your perception of the world we presume to share a common view

You can only say certain things to certain people

Drunkenness is degrading when you lose your faculties

Women can be submissive wanting to be protected

Reality is what you know in weight weathering any pressure

Nationalism imagines a copacetic union

Cable television is demographically mis apportioned

Movies impression the world has many perversions and it does

Notes:

Section Eighty Eight

Metalling minds ionize relations

Paternalism has any emotional confliction

Honesty has many fronts

Walking after meals helps any digestive tract function

"Memorizing" historic figures has no intuitive juncture

You know anyone by how comfortable you are in first person

A world view is a media cultural divergence

Driven people are much consumed wanting to be successful

The family malady is a paternal antipathy knowing what anyone has been through

Chewing foods well helps digestion

Character assumes a lingering posture

Any appreciation of another's time is well spent

Self-knowledge has a positive affirmation

Any law according freedom isn't one's own choice it's suggestive

Government representatives should be required to have sociology degrees

Alcohol lessens inhibitions at times a bit too much letting it all out what you truly want anyone to know

Freedom is where you have no secrets

Time is the obligation to a set parameter

There is no future only the now in any physics to change

The arts express sentimental commonalities

Having goals conditions perspective

In general film depiction is backwards being social aptitude has been evolving

Have you ever been awake in a dream knowing what you were doing

Strength and bravery are conditioned by a physical and mental aspiration

Ladies leave them eyebrows alone and keep any make-up simple

Notes:

Section Eighty Nine

Conversations trigger anticipatory responses

Silver spoons make people different who don't have to work for it

Laziness is a domesticated dog's life waiting for a meal

Songs enchant melancholy thought

What has any hand in your wallet

Social personas impression minds

Violence is an act to insecure fervor

Hollywood doesn't know us to well, does it?

Consciousness has no objective

Ontology is the question of being

Grief is when nothing good seems to finds its way

The sympathy card is a stacked deck in seduction

Posterious dominion is a maladroit social compatibility

Humility' the acknowledgment to your proceedings lacked credibility

The culture slowly grows upon you until you finally realize nothing is new

Paranoiac senses entreaty social internship

Marital domain has a maladroit paternal obligation

Most conversation concerns what's going on with people you know

Movies reflect changing modes improvising social dogma

Social views climb Jacks Bean Stock searching for the goose who laid those golden eggs

Having money affords life style

In youth having no incom heft is an obscure nuance eyeing others adornments

Economic shortfalls initiate intuitive revivals

You want to connect to people with your intuitive stinger

There is nothing honorable to judicial subjectivity

Notes:

Section Ninety

Thoughts not aligning with anything just happen that way

Television violence impressions outlandish human perversions

Regarding movie portrayals terrorist scenarios initiate cultural prejudice

Intelligence questions anything that did not make sense

Patriarch chameleon is a drone's yard

Citizenship is an idiosyncratic ornamentation

Insecurity is an idiosyncratic distraction

Recessions have isolated segments

Con artists play off the idea you are naïve and you are

Writing instructions follow any puzzles procedures

Pride is an auspicious conjecture

Reality is one's own cultural aura in reflection

The malady; it's a fog you know when you are stoned having no connection to anything

Perspective is any biased prospectus

Family and friends is the common social venue

Beware the empathetic emotional loom

The illusionist wants to control your mind

Imagination takes us places on one path or another

Over excitement is a grandiose adventure

Know any facts drawing any conclusions

Complacency stirs emotional futility

Celebrity commentators plagiarize journalism

Having no money as a child theft is a mysterious temptation in development to conscious objectives

Any "chip off the old block" is suggestive in genetics cloning personal traits

Love is trust and understanding

Notes:

Section Ninety One

Opinions are suppositional conquests

What conditions perspectives is positions to industrial economic bias

In medium transfer' video graphic design obsolesce physical arts

Rural flight there is little to no industrial expansion in small towns

Sympathy seduces emancipatory appeals

Science reaches any limits to physical discoveries

There is what you know and what you don't know about people

Any religion is a personal philosophy martyrs and alike

Any social sense is any atmospheric pressure

Infatuation trains congestive thought

911 was an all stop moment, were you there'

In dogma you will always have good memory being consistent knowing what you do

All intuitive art knows the same culture

In procreation it is instinctive seeds your genetic future

Accumulative household memorability is a maladroit trunk

Making connections following any schedule conditions thought

In robotics technologies are more depreciating than human craft

What people do instigating trouble stimulates a temporal caution

Your external presence is not your internal thought

Paternal assumes an obligation postulating scholarship

The world is a menagerie in accumulative discovery intensifying educational venues

Law is a maladroit jungle

Material sustainability is a corporal leash

The world view includes many fanatical theories

Sodium is a drug contributing to stress and insomnia

Section Ninety Two

Truth trumps pawn

Withdrawals condition stress causing involuntary craving compulsions

There's no shame in life only what you do to yourself not living up to your own expectations

Confronting someone setting anything straight involves duress

Love posturing acceptance can be arduous

Drama lures wanting to resolve any situation

There are three layers to life what you know what you do and what you are a part

Expectation lives up to any traditions and health concerns

Familiar settings condition any aura

Any ideological convention is a maladroit supposition

Eighteen imposes consensual age is maturity is an estranging dogma

Mortality is a real fear at an early age; you don't know how to talk about

Cows don't want to be your dairy slaves, do they' look um in the eye'

Exploration in discovery keeps any mind in focus

Youth is impulsive perusing the opposite gender; you'll be better for it later having financial dowry

Eggs are embryos not dairy

A driver's license conditions forage propentence

Can municipalities suspend your license for vehicular fines in default

Over bearing parenting infer they are afraid you will be damaged for life

The price of gas conditions the pulse to expendable income

Being too generous is a posterious invitation to dependence

The brain does not condition the mind the mind conditions the brain

Movies impression stunts children like to play dare devils

There is what you know and what you don't know about people

Coffee is an acidic carcinogen stimulating artificial energy inducing anxiety in withdrawal

The fact that the universe exists is mystical

Notes:

Section Ninety Three

The social view has many perversions in how people's lives are degraded to squalor

Civil rights are suggestive where there is no control over culture

Have you ever looked stupid to yourself not knowing how to respond to people

Motive and incentive are isometric complexes

If you don't want any body fat you will not eat meat

Peter Pan' Captain Hook, Lost Boys' transgression has a nefarious fraternity

Since alcohol it is so bad for us why it is legal' law is bias

Police are people subjected to procedural protocols

You want to be with someone that makes sense to share your life with

I don't have all my faculties together right now I don't know who you want from me

Forage propentence transfer in economy has a natural right in equal origin tenures socialist ethics

When nothing works out people simply do what they do to survive

Media negativity stigmatizes any view in memorialist fashion

There is no such thing as intelligent design knows where anything goes

Crazy is when you think you know anything

Perversion is a mystical accessory

In western material culture it's a natural right to have an income opportunity warranty

Oppression is the lack of a collective cause in the force to conformity in symbolic unions

Law has certain conformity to an illusory view of what "obedient submission"

Regarding child support delinquency does a state have the right to suspend your driver's license sequestering mobility

Finding a mate is inhibiting unless you wait for the one that fits your own style

Playing hard to get with the one you want initiates something of control ploys dominion

The inner part of you' lives in a shell of isolation passing time in illusion

You are what you know; nothing about a collective conscious mind of any authority

Perfection respects common knowns to any social nuance

Notes:

Section Ninety Four

Media impressions copacetic ideas

Society worships successful people

Media trains a pragmatic conscious perspective

No worries' having no goal does not mean no adventure

Concerning liberty authority convicts dominion is a posterious fold

All humans have a forage propensity to survive conditioned to an agricultural apparition

Natural law says you have a right to nature

Social arrogance is a flattering insult

Why do people say stupid things that don't make sense

Liberty is an adulterated child

Paternal domestication provides financial material sanctuary

Corporate advertisements impression good will promotes misleading domains

Conscious reality in observance is an existential posture

There is no such thing as any control is there

Materialism is a preconditioned directive conditioned by culture

You don't know what anyone knows you can't communicate everything at once cause's recipient anxiety

Focus has anything to do with knowing what you are doing

Memory has any idiosyncratic wardrobe

Where is government in any equilibrium to equality

Reality is the physics to nature

Involving constitutions there is no intelligent design there is no science to culture knowing any cause

Children demand structure fostering safe haven

Reality is any cognitive relevance

In youth environment conditions social portrayals to defensive origins

How you are is who you are with anyone

Notes:

Section Ninety Five

There is no such thing as changing the world how does one change consumerism

You do not know anything true unless you know what you know

You want to be with someone who compliments personal interests

Discerning addiction smoking conditions motor memory compulsions perching lips and drawing can only be broken by starvation

Conspiracy regards what government does not making sense posturing any character association

The little voice in mind is the voice of reason balancing treason

Home equity provides a nest egg sanctuary for retirement taking care of things

A world view is an accumulation to experience is a discriminatory perspective

People gravitate to science to make physical discoveries

In transient causes global views digress what government's do in police state temperaments

Memorialization's source national bias predispositions has any conforming inference

A social draw has a lure postulating associations to being anyone

Authoritative cogent coerces submission imposing obedience is incriminating

There's no collective conscious mind everyone has their own signature

Memorialized nationalist travesties are daunting consoles

An overabundance of wealth extinguishes material reward

The law is a transfer in vengeance instigating an authoritarian specter

No one knows your own portrayal to time there is no peripheral recording

Subjectivity to social perversions rouses discriminating appraisals

In youth permanent records nullify any sense to emancipatory causes

Authoritarian; you do not know how to fix people living up to your own expectations

The shadow; Law has a mysterious social aura

Municipal codes initiate contractual deals with coerced signatures

There is truth in philosophy regarding freedom to expression defines common interests

Was Jesus crucified for educating people on our culture and by what authority? Who' ever died for our sins, any mind is an individual responsibility...

Notes:

Section Ninety Six

If you do not want to be bound to someone for life don't have children

Energy and environmental concerns always reappear during economic down turns

There is no comprehension to anything non existent

Existential dualism is any posterity to idiosyncratic persona's

Alcohol is not a glamorous experience; it's a legal poison causing a mental disorder

Regarding suggestive influences alcoholism is not a disease and has no genetic heredity

Marriage is not a law' companionship is a paternal commitment

Growing up in poverty instills an impression imposing characterization

Games which apply any construal practice organizational skill

You don't want to over compensate anything with children they want to know how to know anything on their own

Regarding surety qualification trade schools provide efficiency

For many there is no place to go limited in exposure, education and opportunity

Materialism has an existential personification

Movies impression anyone can make any accounting postulating a true story

Regarding rights there are none you earn any provisional sanctuary

Love is a reciprocal appreciation

Movies concerning high school often imposter norm to social dialog

Garnishing loyalties infers likeness posing any stance

Concerning any physical supposition any investigative construal deduces fact

Being good at anything requires practice even being personable

Regarding and all-knowing spectrum media impersonates social nuance

War movies ionize views imposter national characters traits

Judicial restitution assumes a social principality

Isn't most music paternal lamenting

The universe is infinite is a mind boggling analytic mystery

Notes:

Section Ninety Seven

Work schedule conditions mind consuming time

Nothing makes sense when no one knows what they are theorizing

Debt instills a confidence honoring an obligation fulfilling material desires

Competition in people is egocentric

There is no social congruent karma energy field posturing idiosyncrasy

Disciplinary action cogence lures any rebellious antipathy

Law is non-emancipatory knowing casualness, law is not science it is reactionary

Seriousness is a mysterious disorder

A social view is any idiosyncratic perception in comparison

News media impressions much negativity

Marriage law contractualizes loyalty

Cop shows are a social perversion

There is no such thing as a true world view

There is nothing to know there is no matrix code we simply exist there is no architect only human ingenuity

Speculation has an ardent guard in social deliberations

Playing hard to get she's volatile

Does anything ever go the way rehearsed

Regarding social nuance tone taunts ego postulating stature

Revenge is any auspicious nuance

Any social stage can be inhibiting

Cursive slang has no conscious construal

Regarding likeness protective roles impose to own you or anyone

Main stream media caters celebrity envy

You feel stupid when you think there is anything to know culture is what you know

Elephants are not afraid of mice' who came up with that idea

Section Ninety Eight

Regarding the mind innuendo twists the tourniquet sequestering interjection

When you are drifting into thoughts you are memorializing perfection

The fact that the universe exists is mystical

Reality is what the means does

Regarding economy there is no independence

Paternal senses are always looking over the shoulder to what law lures

Dominance has many stratums posturing what one purchased

Being humans there is the worker bee and the social bee sighting venue

You cannot experience what goes on in a movie you can only fraternize dream states

Image persona is a social apparition

Space is not a physical property it does not curve

Never believe what anyone tells you know what you know on your own

There is no true world view only what you know

Television series mimic culture

Being smart is not being involved with anything that doesn't fit your own character

Movies impression time is a changing fashion

The media conditions the view impressioning social effort

Dominion and necessity condition views to an interdependent economic conditioning

Memorializing time a world holographic age is any change in fashions

Good acting roles chameleon approval

Unemployment figures do not count everyone specifically high school graduates who have never worked or construction workers who are self-employed and have no contracts

Postulating restitution there is no criminal intent hitting a telephone pole siting felony

Corporations are not human entities postulating entitlement rights to personal property

Corporations have no emotional attachment to existential properties

Talking about people in a derogatory way is ostentatious

Notes:

Section Ninety Nine

In education nationalist associations are memorialized transient causes

Killing anything poses a conscious decision

Approval personates individual character mirroring attraction

Regarding reproach law insinuates one mind postulating authority

There is no national one character anywhere

The law induces temporal currents conjuring suspicions

Regarding honesty and esteem failure and practice is synonymous

Modern convenience makes lazy people

Good lyrics tell true stories

Regarding stature there is nothing to know there is no circle unless imagined

Corporation's condition views impressing monolithic structures

Aggressions initialize consuming pedigrees

Pessimism limits dialog

Being cornered vacillates responsive retort

Municipal ORS codes in general opinionate pragmatic correctness

Media impressions technological sophistications

Regarding opinions there is no expertise

Attentiveness is influential inspiring loyalties

When one takes responsibility for anything one assumes an obligation

Thinking anything through is a pragmatic definition

Kids are kids born into so much stimulation can be consuming in what to know

Commitment is a compliment someone worked things out with you

Concerning subjective remarks, anything that doesn't make sense pinches a mind

If you state anything properly you will never have any lingering regret

Intrinsic arts impression common social nuance

Section One Hundred

The panoramic social impression is an inelegant portrait

Immunity is independenceI mind mine mind yours

Authoritarianism; what is ominous expectation

Praising a child caters to talent in appreciation

Posturing social imprudence posturing authority reveres law

Knowing what to say to anyone solicits common associations

Globally we will never settle our differences unless we understand cultures

The primacy to human culture is vanities innovation

Constructing thought reasons contextual portraits

Taking charge of any project applies cognitive organizational skill

Any social sense is the mind in comparison and appraisal

Occupational commitment acclimates time influencing emancipatory goals

Suspicion scrutinizes sincerity

There is what you do what you have to do conflicting with what you want to do

Frustration' it did not work to your perfection

Paternal sense mirrors the aura in expectation and approval to cultural traditions

Cyber space, we have reached a pinnacle to information overload

The paternalist wants to fix you up prognosing uncertainty

Social antipathy is there is no whole truth to anything

Character wants to be known for adventure

Cultural memorabilia is memory using no cognitive construct is ambiguous savvy

Philosophy' there is no union where there is no social gathering

There is no collective mind recording any knowledge to grasp from thin air there is no physical world view only what becomes recorded accurately in retrospect

Status quo assumes a common aptitude what is it

Section One Hundred One

Physical pain conditions temporal fixations

Love is an intimate reciprocal connection

Social maladies are a soap opera's rendition

The mind can be consumed reliving any experiences

Regarding GPS, you don't know anything you haven't mapped out in your own mind

Gun laws don't change transient causes inciting rebellions to protective rights forays

Black market munitions circumvent authoritarian protocols

Unless there is a dialog to follow television and movies are mind numbing

Knowing how to plan a career is an essential part to life knowing what to know'

Especially WWII' war movies seduce youths to militaristic conventions

Regarding maturity marrying early sequesters social commentary

Crazy is assuming there is any intelligent design authoritarianism is auspicious

Intelligence is competitive in historic cultural interpretation wanting to know any cause

A woman's role is not a maid, nor a husband's mother, are submissive dominions

Domesticating roles have been and are merging, home economics class is necessary for anyone

Regarding the law what gives anyone any right to cause any ones heart to race

Nature is harsh not romantic unless you are a sadist

Election polls are suggestive influencing votes is any posterious consult

Staring at a law agent while driving is no probable cause

If you want to keep your mind free don't use drugs

Cannabis is not a drug it is a plant why is there economic laws against growing flowers

Concerning forages propentence economy is falling apart every day for many people

Consciousness is conditioned by culture in adaptation there is no primordial knowledge

It is the existential cultivation which drives economy not government though government is conditioned by cultural hindsight

The inclusion of a child within the simplest fashions of practice stimulates recognition to their presence' lending themselves the esteem to their own sense of being

Notes:

Section One Hundred Two

Video games stimulate sensory motor behavior intermits focus conditions temporal impressions

Computer keyboard word processing macros obsolesce hand writing

Running out of money is a degrading consultation

Time and expectation pressure aspirations

Intangible machines compel intrinsic tools

Most video games carnalize physical aggression

There are dream worlds and there are envisioned worlds to merging polarity

Being lost in the wilderness disorders perspective to existential wardrobe nutrition

Paternal is a posterious polarity in domesticating principality

Involving pressures providing for family conditions conscious subjectivity

Cultural embodiment is a corporal collective experience

Authoritarianism is primordial extortion

The primacy in competiveness to knowledge concerns anything known

Good grammar has any definitive physics to cultural epitomes

When life becomes all about the money there is less time for the artoit

When you think you have life all figured out' think again and again and again

Regarding thought any regret may resurface many times is congestive

Pending matters conjure thought conjuring conclusions

High school impressions character references

Fatigue causes irritability

The preacher's quire doesn't change a thing

Movie drama induces emotional paternal impulses

Mating is a primal instinct in reproduction

Environment is incriminating

Physical mediums develop into robotics obsolescing craft

Section One Hundred Three

Debt is acquiring what you have not yet earned

A cultural reflection is a changing perception in familiarity and exploration

Living with someone conditions time and accountability

Naivety has a vulnerable trust

Most relationships are superficial masked with deception

The social view is a class menagerie

Environments condition manner in adaptation

Occupation wants a separation to time in temporal dominion

The social malady is an economic disposition

Any world view is an economic division

Character has and aspiration to be known for integrity

Mummy movies aren't what they used to be

Parents think children know what they know about the world is obvious assuming a preconditioned intuit

Any world view perspective is an idiosyncratic wardrobe in dispositional attraction

Hardened men fashion harden women

Male insecurity and dejection (mom wasn't there) lends to unworthiness finding a life companion

Life and health healthy choices are a persevering nature

Social manipulation is a leper's quarrel to undermine to dominate exposes insecurities

Not knowing how to respond to people who are manipulative scrambles any mind of reason

Memorializing anything throws anchor astern

Concerning taxation you are coerced as a prisoner and slave to a material extremity

Physical humans medium extensions in posterious taxation knows no unified cause

Defeat can only be achieved thinking there is any end goal

If there is no directive there is no practical course

Where there is no intent there is no concern doing what you think is right in what you do

Notes:

Section One Hundred Four

Life is always reinventing the wheel to a journey

Blending flavors is a culinary art

Life's dreams are a journeys wake

Commitment has a mutual appreciation to life's impediments

Social envy postulates an inner and outer circle in social status to a paternal fraternity

Leaving a perch customs any idiosyncrasy to impressions seduced by the desires to patriarchal attractions

Love is not what you think it's what you know in confidentiality

Having dependents extends a posterious aura being residual

Interpreting law is anyone a ward to state obsolescing liberty posturing authority

Not having anything constructive to do and addictive tendencies is synonymous

Independence is not being conditioned to anything

Ostracizing any faults conjures rehearsals

The law judiciaries' demeanor

Television comedy series parodying vindictive assertions is non intuitive

Regarding media impressions you wouldn't know anything you had not been subjected to

Memory canvas's any quality to life what does anyone distinguish in ornamentation

Social innuendo plants seeds in suspicion

Metamorphosing features in movies leave lasting impressions on children

Media devises are envious attractions for youths looking cool

Undertones relating involvements solicit contribution

Thinking anything through is an intuitive proclamation

Money has a constant necessity conditioning a sense in survival

You don't know anything you don't know anyone tells you in any opinions of other people

There is no collective state of mind only individual dispositions

No child has an instinctive tendency to comprehend the relevancy to our culture in forage transfer

Notes:

Section One Hundred Five

Invention is a human envy

Freedom is entrepreneurial expedition

Humility in ego retraction knows no hold

Speculation is an intuitive consultant

Retaliatory causes are the people's law

Restitution is a mercies plea bargain

Infidelity is dejections reprisal

Extreme situations muster fervent courage

Youth has certain optimism

Quality speaker's rock

Vehement expression has a loss of innocence

Sexy is too hot to handle don't touch this stuff

Donate any unwanted musical instruments to schools is a good cause

Film impressions life as a motion picture to time

Does what you have make you who you are

Empathy has an intuitive reflection in seduction to uncertainties

No one knows your own perception of reality

Perfection is a sign of insecurity

Sincerity has a certain tone

It's a mystery what any people around you know

Marriage is a tug of war in loyalty and trust

Arts in music are a selective ambiance

Poverty is a very degrading malady

Saviors are apparitional guardians

High School summer vacation had a sense of freedom

Notes:

Section One Hundred Six

Capitalism is twenty four seven

Malevolent people provocateur pickets

In social hazard disability payments should match cost to living standards

Embarrassment has many relations

Impoverished people are more real

Economic fall backs are awkward havens

Deceptions have no foundations

Adam & Eve physical awareness is an existential menagerie culture conditions conscious objectivity

The presence of age and being is a portal in world composure

Physical attributes condition environment

Ego is a scholarly observer

Death and war in youth has a gloomy aire

They built Pyramids in the desert to rise above the planes

You're not stupid there is no intelligent design

Technology fashions the perpetuity of change

Deception manipulation conspiracy is a fool's bargain

Love is conditional you don't turn the light off to turn the light on

Adolescent economic transition has a rookie's pardon

Life is a conversation in mind

Childhood dreams are greatly diminished by the pressures in life's occupation

Plea bargaining is an erroneous extortion

Government is doctrine having no mind subject to transient interpretation

Notes:

Section One Hundred Seven

Red tape doctrine is proscribed knowledge on any applicable basis

The only way to beat addiction is starving physical memory

Pretended knowledge in conversation congests the mind in rhetoric

To "Believe" is having faith in what you do not know

Favoritism and Jealousy conditions sibling rivalries

To believe is not to know anything imagined

Words associate memory to functions

Social conflicts in youth are disturbing numbing the senses

Short term memory is conditioned by routine

Mindless debates where there is no context causes consternation

Memorialization is the main cause to human conflict

The rise in autism and **ADHD** is relative to social economic pressures

Living in poverty having no money or opportunity causes social antipathy

Social transgression conditions suspicious thought postulating motive

Transgressive posture has an anarchistic plea for social definition

Saying things in derogatory despondence embellishes no mediation

When someone is lying they become animated conditions social antipathy

Admonishing personal character in unwarranted accusations in social conflicts with deceptive respondents is operose

No contact orders limit consummations to separate version disseminating fact and intent to social provocation

Science to law; defining conscious entity

Media devises in social community is a consuming addiction

Expectation knows no cause conforming childhood lucidity and imagination

Love has a certain charm until you live with someone fostering etiquette

Nothing is as it seems in attraction until you have lived it realizing it was envious emotion

Societal comparatives are very estranging in youth

Notes:

Section One Hundred Eight

The distance in her eyes was an appeal from the depth in her desire

Reasoning anything has any sense in resolve or achievement

There is no intelligent design there is no cause, consciousness is conditioned to culture, and drugs take you nowhere thinking there is anything to know

Addiction recovery is most conducive leaving environmental familiarity has a change to mind

You take responsibility knowing the conspiracy, any facts to be presented and weighed by a jury of peers

Thought in cultural perversion has any ardent retort to association conditioning a reprobate mind

We all have knowledge by the perversions to our culture conditions memory

Nothing is as it ever was before' there is no collective mind

Television shows displaying social perversion have lasting impressions

Music has any emotional release

Any musical art has its own original distinction

Anxious people condition ambiance

Rock music has many modes of expression to social distortion

Quantum physics is the cultures epitome

All memory has contextual prompts

Fear is conditioned by any desires to be known as being successful in any force in expectation

There is no god of knowledge in architecture to an evolving culture

Every individual language has its own personality

No one knows your world view

Attention span in small children is mystical menagerie frolic

Child molestation preys upon the naivety of innocence

Adaptation in malevolence of youth sensing exclusion in some limits maturity wanting to be engaged with a past innocence

If you were molested as a child there was no fault there is no shame

Extravaganza fashion impressions much envy wanting to be wealthy

Mystic minded people are martyring harps

Notes:

Section One Hundred Nine

Human interaction has multiple versions apportioning intent

Character has a familiar hold

Pursuing an art verses a career is anti-authoritarianism

Intelligence has always evolved in technologies sophistication

Transgression is a virtuous trial

It is very difficult to put experience to words knowing how to respond where there is no truant dialog

Reality is physical existence

Don't ever negotiate with terrorism

Lowest price Lasik eye surgery isn't the best value

Who you know has any idiosyncratic past

There are no absolute stories only what culture does

Paternal senses foster any resolve

Inflationary bubbles call for better constitutions

Addiction conditions motor memory

A true world view is a metaphysical consult

Intelligence knows memory relativity

Anxiety knows any expectations to tenacities

Sexually favoring same gender is a domestication inhibition

Recycling should have unified global codes under corporate management

Consciousness can experience any mystical psychosis using Cannabis trying to comprehend culture defining reason

Using cannabis can cause a keen sense of self-awareness "paranoia" postulating polarity exposure

Nationalism conditions conformed world views

Fames presumes to know a world portrait in motion

Authoritarian nose is a bureaucratic warden

Carcinogens conditioning the tongue are pacified by stimulations

Notes:

Section One Hundred Ten

When you have confidence you know how to handle yourself with people

Nature calls' Breast feeding in public is not a social perversion

Given a choice between poverties depravation verses wealth's luxuries anyone would choose the ladder

Economy is not an institution challenges government roles

Sharing yourself with someone sexually has a paternal possession

Materialism has an accumulative residency

Worshipping idols has no intuitive virtue

If there is nothing there' there is no context

Facebook can be a consuming attraction

Catering to aggressive animosity has a sense of being controlled

Codependent puppet strings are conditioned by retaliatory dejection

Humility is the cure for anything

Leaving home and community for abroad has an estranging paradigm

Money provides safe haven

There is no physical time' time is a cultural condition

Nothing feels like anything imagined

With people' you know what you know when you know what you know

Who wrote the bible is a massive riddle

Department of Homeland Security is communistic

National news is corporate monopoly

How does anyone describe an aging perception of the world

Knowledge to culture is the only authority

Television impressions people are stupid

Transgression emits social distortions has any residency

The strife for liberties preservation is the restitution to transient authoritarian despotisms

Section One Hundred Eleven

Social traumas induce psychosis (deep transgression challenges intelligence) reflects a bi-polar dejection in self-value stimulates fantasy obsessions based on paternal desires for perfection living up to traditional expectations

Existential dualism is the true psychology

Speculation is an intuitive characteristic in cultural survey

Media impressions a pretentious aire

A police state is a militaristic provocation caused by economic trepidations inciting intellectual revivals

Heroine fantasy wants to take the stage and change the world

The sentiment of exclusion from family is an estranging paradigm

No one knows anyone's world view

Cultural speculations are foggy spectrums like multi piece puzzles

Metaphysics is cultural linguistics

The world is superficial unless you are involved

There should be competent test for voter registration

Being questions what to know the answer is your senses

Socialism converts to communism has an economic cycle in despotism incites rebellions

Societal resentments stimulate retaliatory thoughts

Police using semi-automatic weapons combined with anxiety contributes to law enforcement fatalities

Age has awareness to laws accumulation sensing a loss to liberty

What are any parameters to normal in psycho analysis

Much "*institutional*" knowledge is on a need to know basis

Random thought has any mysterious appointment

Low value senses dejection incites hatred and resentments relates to sexual abuse crimes

Gymnast ribbon Paton's are a great toy for kids

If you are not inhibited you can make millions acting stupid as a comedian

Being told what and how to do anything challenges competence

Positions in authority construe contested procedural efficiency

Notes:

Section One Hundred Twelve

Having children *"involves"* local community

Ideology rationalizes idiosyncratic virtues

No one is ever who you knew

There is no meaning or purpose to life there is nothing to know

Manipulation always has a twist in lure

Self-evident; if you have no knowledge in culture you don't know anything

There are things you know and never know how to describe

The greatest desire wants life to make sense

Conversation in mind is a sounding board to reason

Assertive compensation caused by insecurity has a princess court

National corporate media has any bias world view

Any measure has a sense to subservience or supremacy to any derogatory connotation

Grading should be focused upon systematic functions

911 was not carried out by Muslim Religion is was fanatical terrorists

Family environment conditions subservience

Speculation provides scenarios to motive

Competence has an estranging association

Much speculation cautions outcome

Culture is conditioned by human ingenuity not by imaginary gods

There is no physical motion picture to time no one knows your perception

Forage propentence transfer in domestication has many tolls

The conquest in betroval has a luring challenge bridled to marriage

Is it cool to simply be in fashion

Being real has any attraction

Tradition has a social paradigm

Notes:

Section One Hundred Thirteen

Royalty has an ostentatious pomposity

Self-deception admits no fault

Thought is conditioned by patterns in familiarity

Doubt massages no Aladdin's Lamp

Human intelligence in general has been evolving

Hygiene conditions awareness in human vanity

Indigenous people are more uninform in appearance has less distortion

There is no debt in obligation there is only your own personal conduct

You can't judge a world's norm by common media impressions

In nature's dogma' the human mind is a conscious entity to one's own domain

Modernization distortions Cottage Ville pleasantries

Manipulations always have a positive lure buyer beware'

War makes angry children terrorists so kill the babies while you're at it

Authoritarian law conditions an existential station

Those working construction are not given enough appreciation in the strife of their labors in the transience to opportunity

What do you talk about you don't know'

Facetious behaviors intended to make others look stupid is extremely infantile

Authoritarian is a repressive thumb

You are what you know' materialist is the existential human condition'

There is no context to any purpose conditions conscious objectivity also stimulates cosmological theory in suppositions to an alternate dimension of the universe

A singularity of provider can lend to animosity of a spousal dependent sharing no financial burden does condition subservience

There is no intelligent design, consciousness has no pre-conceptional operativeness

Oppression is a monetary degradation to natural principles of forage conditioned by agricultural despotism

There is no intelligent design there is no Neo or the One'

Typing efficiently using a key board is "trained" second nature memory not learning

Notes:

Section One Hundred Fourteen

Truth has any residence to social resolves

Social rage/child abuse is related to roles in responsibility in limited communicative intuits

An individual moral postulates a clear conscience

Denied a sexual experience in an extended relationship has a sense of dejection

Trust has a domesticating paternal commitment

Social tolerance has a naïve warden

Insolence preys upon naivety

It is easier to resist any temptation in the mind by thinking anything through than performing the action itself losing self-discipline

Academic tiers have eclectic piers

Academic stature has emotional desires postulating virtuous standards

Academic scholarship in grade school wants linguistic tools

Story telling is a practiced art

Having a confiding relationship with children sets precedence in trust

Presidents have magnetic armor

Commitments want special interests

Educational commitment wants an ardent blue print

Population is a true concern calls for better education

Mathew VI 22-34 there are always two of you the observer and the observed checking balance

There is no one mind there is no collective mind knowing cause

Perseverance wants a mark on the world

Royalty knows no cause

Conscious relativity is a domesticated culture

Family planning in poverty could consider subsidy life incomes for fertility sterilizations

Racism is a non-intuitive anthropology

Idiosyncratic trivial knowledge has any national bias

Section One Hundred Fifteen

Free trade capitalism knows no inalienable rights

Nothing is as nothing was

Parental paternal empathy is arduous in cultural deliberations

Congestive thought knows no story

 If there were any goal in life it would be pursuit to happiness

It's all downhill after you get married unless you are in love

Capitalisms perversion mandates mental health funding

Mind is an environmental impression conditions character

Culture conditions governance there are no constitutional blue prints

Listening to others childhood stories often center on authoritarian abuses

Fast for three days before dieting

Reality is physical existence

Ageing has an appointment with the unknown, be optimistic you can never know death

A world view is the attachment to time in changing fashions

Speculation condenses cultural fog to fluidity

Nothing comes from nothing

Inclusion postulates a common view

When you do anything that seems wrong has a sinking aura

Group fasting sabbaticals

A common story is a social soap opera

When you make sense to anyone a light comes on

Life has a pull, where are you going in such a hurry

Meditation, fasting, isolation curbs all stimulation

The moment is the future

Denial assumes a cause

Notes:

Section One Hundred Sixteen

The real world is a masquerade

Erections are excruciatingly painful during puberty masturbation is a healthy choice

You can never presume another person's confidence

Social mockery is a fool's parody in empathetic subterfuge

If you have been charged in a civil matter and are not guilty go to trial and no one will show up to testify because they would be lying

Television sitcoms lampoon social norm

Enjoying what you do for a living has a sense of value

Transgression has a taunting seduction

Intrinsic art is a sub terrarium periscope

Honesty has a twisted convenience in retrospect

Running out of things to talk about people make things up

Six of one half dozen of another nothing changes

Sexual arousal causes much perversion in thought

Poverty has an oppressive character

Comprehension is easier listening than reading

Social transgression is all suspicion

Corporations controlling forage commodity transfers feigns social responsibility

Military isn't the worst program' age 18 nowhere to go

Knowledge provides confirmations gains confidence in what one knows as non-verbal

Rationality weights options to outcome

Jealousy is a maladroit perversion

Transgression is an economic malady

Ego wants approval has any invention to paternal deities

Neuro Science; perception to orientation "awareness" is all frontal lobe

Social subjugation venues Macomb spectacles

Notes:

Section One Hundred Seventeen

Every nation has its own idiosyncratic impression

No one knows you unless you reach the top of a celebrity parade

Money is the deepest existential forage transfer

Postulating a collective consciousness assumes common knowledge following a plan, intelligent design is human ingenuity

Discipline has any follow thru to goals

The force in culture occupation knows no freedom is a bipolar influence

Preceptual operativeness is instinctive adaptation to culture

Limiting perishable foods on hand you will not be pressured to eat as much

You cannot dispute nature's truths

Love is not a codependent emotional union love is intuit

Hydrogen and Biofuels are the answer to meet the needs of renewable energy resources

If there was anything to know in religion there would be common knowledge, there is no design there is only evolution' including intelligence'

Terrorism can only be defeated by knowledge that all cultures share the same origins

Selective historic impressions educated in culture should not be graded'

Appreciation companion's unfulfilled desires

Nicotine absorbs in the tongue conditions addiction withdrawal; 3 days fasting works

The King James Version of what "Cultural anthropology' or Philosophic Hermeneutics

Innovation has always provided efficiency to meet the demands of a growing populace

Belief in a God imagines a deity knowing one's own conscious life experience

The belief in God wants to make sense of life postulating intelligent design follows a plan

A Dictionary is cultural conscious vocabulary

Authoritarian sober postulates obedient oppression

Authoritarian assumes a collective mind knows all laws' paternal law knows common senses

The drug Lsd does not enlighten a mind with knowledge'

There is no matrix collective conscious view

Notes:

Section One Hundred Eighteen

Seduction has a lonely pout

There is no intelligent design has any appreciation to science

Normal is a material drone

Scripted knowledge has no consult

The world is so massive is confusing challenging what to know

Nothing is until it was

Bipolar is a codependent martyr

The health of a nation is weighted by its crime

I Q = Contextual relativity

A teacher is a good student

Asking; how are you" is a posterious lure

You can never assume a child understands all to what you say answering query with questions

Reading combining visual graphics accelerates comprehension in functions to contextual relativity

Training children to material cognizance has equivalence to placing a saddle on a dog

Poverty has an oppressive aura

Authoritarian reverence is an ambiguous council

Complacency & depression is the leading cause to obesity

Infant transition to childhood has an estranging doctrinarian on innocence in the obscurities to authoritarian subservience

Assuming authority in education for the health and welfare of a nation's people government shoulders a responsibility for any mental illness fashioned by the cultures perversions

Socialism has an inelegant custodian

Youth has a posterious fashionability

Any collective world view is any ardent outcome for resolves assumes a common mind in knowledge

Desire for attraction charms heroine mesmerism

Conscience has its own liability

Religion is ones moral dogma

Section One Hundred Nineteen

Jesus metaphors martyring artists plagiarizes the science to philosophic hermeneutics

Embarrassment has many character inferences

Respect has an earned appreciation

Resisting Arrest; resisting captivity of the law is instinctive

Cycles to economic despotism caused by recession cause patterns to police state communism

In the case of individual property corporations have no civil personage

Western involvement in the Middle East since 1991 provoked Islamic terrorism

Public Corporations are a form of communal commodity property is pseudo communism

Everyone has the right to subsistence in forage propentence transfer to material domestication

There is no collective consciousness there is no mind in knowledge only common senses

Human complexity challenges an evolving intelligence in social interface

The cause in democracy is preservation to "civil" liberties absent definitions to despotisms derivations

Indigenous origins in temperate zone condition culture, organic genetics and skin pigmentations

The magic to memorization is interest

Population increase to economic growth ratio has demographic disproportions

Arrogance is a fool's pride

Ego has a stealthy front

Complacence is a connoisseur's buffet

Material sanctuary is a serendipitous challenge

Transgression is the condition to conscious objectivity

Malevolent oppression is a fool's envy

Baby talk condescendence postulates incompetence assuming preceptual operativeness to culture

No one wants to know you when you have no money

The will of god is what nature conditions

Religious speculations have no physical comprehension

Notes:

Section One Hundred Twenty

Hot fluids condition cravings in healing withdrawal

Existential is the material epitome which conditions being

Appearance more than anything as we age changes the way we think

Nothing derives from nothing

Any chemical imbalance is conditioned by sociological stressors

Male femininity in drag is a protective disposition to virile transgression

Sibling incest would be related to carnal compulsions during puberty in dysfunctional environments

Youth respects intuitive authorities

Social characterizations during youth is a Vincent Price feature

Truth is the only justice

Greed is a social economic disproportion

I love you has a reserved commitment presuming who knows you

Seduction has a strange way with illicit conducts departing dogma

Denial has a transparent cloak

Social pretention has a superficial aura

What economy does conditions life

Bipolar wants to justify its own distortion in social resentments

911 exasperated cultural prejudice

There should be fail safe programs for recessions and social discriminations

There is no intelligent design asks the question how not why

Reading any mind has common predictabilities

Capitalism is extremely degrading though still provides a sense of freedom as economic slaves

If you are young and live in a small town consider what cities have to offer

Much education is a scripted idiosyncratic menagerie

Sexual trials during youth has certain innocence in culpability

Notes:

Section One Hundred Twenty One

If you think yourself anything you will be what you think' think well

Homosexuality wants to justify its perversion post dejection to domesticating traditions

The main theme to the New Testament is social comprehension being one mind confirming common infirmities

Intellectual meetings are the best social venue sharing concerns and experiences in person

Material interdependence conditions a paternal fraternity

Practice in comprehension keeps any mind in tune

Adaptation to material domestication conditions cultural literacy

Paternal sensuality has natural boundaries in corporal reservations

Much speculation has an incriminating affect in value

You can never comprehend anything anyone has gone through by what they say is still one's own version

Poverty has a transgressive aura

Government positions should require sociology degrees

Over population has always been the catalyst to tribal warfare

Drug addiction is a social illness and should not be treated as a crime

Reading out loud in class rooms is inhibiting

A war on drugs is a communist succession no one has the right to tell you what you can do to your own property

Drug related property seizure by authorities is still a violation of individual property rights

A world view is conditioned by an, individual sense to prosperity

Male dowry is chastine subservience

Financial independence is the cause to woman's liberation

Adolescent rage is fueled by a sense of dejection in value is "transgression"

Culture is what conditions human existence not government

A world view is a paradigm gauge in perception to culture

Adolescent perception is a proprietary distortion

Capitalism is all theft from the same source of human vanity

Notes:

Section One Hundred Twenty Two

There is no preceptual operativeness "intelligent design" to culture the brain has no intuitive function brain activity is sensory perception conditioning coordination

What would you think about living in a prison cell 24/7

It is science not religion there is no preconditioned consciousness to cultural adaptation

Instincts are conditioned by environment and physical attributes for any species

Motive defines poetic distinction

There is no intelligent design authority is knowledge to culture conditions consciousness transfers to philosophy

Transgression is an economic tier conditions prejudice

Delinquent fines for municipal code violations has no authority removing driver's license "privilege" a mobility right

There is the physical reality to existence then there is the metaphysical reality to culture "Quantum Physics"

There is no physical space time continuum physical reality is one multi particle dimension

Reality is an observatory station in the midst to evolving technologies

The code to comprehension is contextual relativity

Conquests in love often end in remorse still wanting to play the field

No one knows any others own motion picture to time

Despotism is a sign of bias prejudice and authoritarian ignorance postulating dominion

Much education is bias idiosyncratic programming postulating a true world view

Children want to be talked to like real people posturing inclusion

Corporations (publicly owned commodities) fall under government regulations in social responsibility

Postulating a collective mind of knowledge has an external polarity

Culture is accumulative congests the mind in memorialization

Appearance has any paradigm prejudice

Democracy in capitalism has a social responsibility to domestic provisions in sustainability

An only child's parent wants a sibling companion check the wallet

Having children affects a social paradigm

Any social view is the people you know

Notes:

Section One Hundred Twenty Three

Listening to others in twelve step meetings practices cultural comprehensions

In forge transfer government guarantees no economic placement to free trade capitalism

The real world is filled with financial hardship

Entrapment soliciting probable cause is malicious authoritarian behavior

Criminal is any intent to cause harm

In existential polarity no one knows what to know... The answer is nothing; we all have knowledge by experience it is not what you know it is how you know anything,

Material domestication has a paternal custodian is the poetical human condition

Human suffering is the memorialization of a mindless culture knowing no cause

There is no collective mind to culture is the reason people invent creationism postulating intelligent design

Intelligence pursues understanding in the natures of culture

Humility is the essence to a happy life

Contagious acceptances in illicit behaviors has a sinister perversion

A polarity in social dejection is the main cause to criminal malice

First time fornication is a very awkward experience where there is no performance

Most crime is drug related (an altered state) questions any faith in religious damnation as a deterrent

Emotional people are codependent martyrs

Police officers have no right to search you or your property without a search warrant or definitive probable cause

Forage propentence transfer to capitalism having no instructions is perverting

You have the right to an attorney before any law enforcement interrogation

There is no collective conscious mind' national media conditions a proprietary view

All media for children should be censored for violence and sexual content including the news

Resentful people like to cause problems in others' lives

The biggest fear in physical altercations is exhaustion, the reason prisoners work out

If we as humans eventually blow ourselves up we would either simply not exist or we end up some sub particle entities looking for our lost cell phones

Capitalisms objective as taught desires fame and fortune

Notes:

Section One Hundred Twenty Four

Much thought stages rehearsals in desired outcomes

Loss in train in thought leaves a mindless stare

Much thought has critical polarity

Making sense in what you say involves intuit

Measuring what you know binds focus in ego polarity

Friendship has a polar persona in appreciation

An extension of the eye and photographic perception cameras obsolesce the craft of painting

Subconscious is a chronicle to cultural experience and idiosyncratic impressions

Mentally Ill homeless people is a sign of political stupidity

Sequestered time for any learning and practice has a nemesis paralysis; Mark IV

Like war on terrorism the war on drugs only caused more human suffering

You have to be a scholar in Cultural Anthropology to interpret the bible

Communism "Authoritarian despotism" is the cause to intellectual revivals (I Corinth XIV)

The basic theme in the teachings of Jesus is one mind however there is no such thing as a collective mind' only individual perspectives

Post 911 War on terrorism only accelerated terrorism in religious fanaticism

The pioneering of America has a transcendent affects to social distortion

Good grades show an interest to any subject

Material domestication conditions expectation

Inflation dilutes equality conditions despotism

Video games condition mindless patterns to over stimulation lending to compulsive behaviors in corporal assimilations

The origins to innovations in human physical medium extension has an exhaustion

The existential world has any idiosyncratic emotional attachment

Existential dualism is psychology "what conditions conscious objectivity" adaptation to culture conditions natural instincts of survival

Surveillance society is an existential authoritarian schnozzle

In the origins to vocabulary language expresses expanding cultural relativities

Notes:

Section One Hundred Twenty Five

A consumed mind accelerates time

Tooth Fairies' children have no comprehension in reasoning believing anything they are told

No one knows where anything goes is the mystery to conscious objectivity

Intelligence knows what you know

Technologies invented time

Drinking water every day helps immunity and metabolism

Soap Operas are transparent parodies

Nothing is as nothing was…

A paradigm shift in consciousness can be caused by extreme changes to environment IE divorce/losing everything

"Culture" Existential is what conditions being in time

Being One Mind (centered) has any sense to resolves in edifying recognitions

Fames seduction has bodacious polarity to fraternities

Alcohol is a carcinogen conditions auto immune affects stress and blood pressure in withdrawal

You feel stupid for anyone acting stupid

Philosophic Hermeneutics is a practice in comprehensions to cultures poetical undertones

Compliments conjure expectations smiling for any ego pose

Playing drums is an athletic physical art

Perspective has any cultural relativity

The theory of relativity is contextual rationality of human conscious existence postulating reason

Innovation in techno efficiency is the cultures ambition

Meth is the most perverting drug

Transgression is poverties child

Caffeine nicotine and any chemical (carcinogen anti bodies) condition auto immunity in pregnancy

Notes:

Section One Hundred Twenty Six

Human culture is non instinctual wants instruction

Language has relativity describing experiences

Literacy has any cultural interest

Intuit is relative to compatibility in cultural conformities

Opinions and innuendo have a certain seduction in retort defending character

Indifference in passive resistance knows no reprobate distortion in reprisal

The brain has no intuitive function the mind in perception conditions coordination

There is no true recording to time only artifacts and monuments postulating significance

Authoritarian obedience has no intuitive juncture

Second hand knowledge has any partisanship

Individual hatred is relative to a prolonged sense of incompatibility testing competence

Democracy has a sense to social liability limiting oppressive tyranny

Hoarding wealth dilutes monetary circulation like a game of monopoly makes unhealthy societies

Culture conditions conscious objectivity in existential being

Cultural literacy is a metaphysical periscope

Approval and dejection has any inclement polarity

Authoritarian postulates subservient as competence

Celebrity Glee club is a serendipitous sorority

Perception is an idiosyncratic symposium of time

The Epistles of Paul allegorizes Plato writing about Socrates his friend

One mind assumes a collective motion picture to time in the same perception to culture

Consciousness is an impression of culture philosophy is a practice in comprehension to what you know

Post High School graduation furthering no education has a sense in custodial abandonment

Pretention has a mysterious echo

There is no collective conscious mind there is no constitution knowing culture, culture is what culture evolves there is no intelligent design there is no authority only interpretation… God is knowledge

Notes:

Section One Hundred Twenty Seven

Fantasy dream states derail any course in realization

Reasoning weights outcome

Mechanical skills practice construal

Individual liberty is moral justice

There is no intelligent design knowing cause' there is no one mind

Commitment has a sense in denial wanting to play the field

Love has no barriers

Cause has no memory

Domestic abuse has a transient aura in repression to social value

Nonverbal has its own language in realization

Language reaches its limits responding to stupidity

Manipulation has a sadistic coil

A chemically altered state of mind has a false sense of invincibility tests any limits

Who you know has any idiosyncratic gesture

The matrix sentinels represent authoritarian transgression

Scripted memory practices no construal in mundane limited structure links to Alzheimer's

Education memorizing people places dates and things practices no intuitive construct

No one is going to save you from feeling unsuccessful

Bullies man up to insecurities

Jealousy has a stirring perversion

"My Love" has a radiant reflection

Who and how you are with anyone is an ambiguous parody

Manipulative undermining wants to hold your dreams ransom

Life has no director's view knowing all the parts

Paul's Epistles allegory Plato's letters (Epistles) writing on virtue society postulating one mind

Notes:

Section One Hundred Twenty Eight

Infatuation in loves desires sensed in dejection coals the flame

A healing tongue knows the most addiction feeding the mouth not an appetite is mostly conditioned by salt

Memory fades in the mist of a new day's dawn

There is no collective mind to resolve

Deceptive people cause many distractions hiding truth to motives

Material domestication in provisional sanctuary offers value in attraction is a consensual age specter

Mind is a conspiracy in appointment

Reality is any individual point of reference has no common dialog

Adolescent sexual experimentation wants to harness an extreme carnal compulsion

Economic placement has no standard cubical

The world is not going anywhere its right there in front of you

Sex is just an orgasm love is intellectual

Unfinished business is a concerning factor

Memories are peep hole photographs to time

Science and religion want to be a part to something that makes sense understanding the nature to existence

Accomplishments protrusion character reference wanting to be known for being successful

Mischievous natures conjure accumulative karmas

Binge eating disorders is a carnal condition caused by physical withdrawal cravings

You will never know what you assume postulating other's affiliations

Belief has no comprehension faith postulates reason

Computer macro procedures condition memory in practice

Maintaining a diary is a ardent of reflection

Most subjectivity has no practiced verbalization

The world is superficial where there is no intellectual gathering

What's going on is all conspiracy in memorialized misconceptions

Notes:

Section One Hundred Twenty Nine

Impurity of heart is a self-inflicted condition in the adulteration of intuitive virtue

If you have a good income your sense of the world is not the same as those who have little to nothing

The movie "Into the Wild" directed by Sean Penn paradigms the wilderness of the philosopher as does the 1968 Prisoner Series by Patrick Mc Goohan

Philosophy educates us to our culture

Grumpy grudges have a peculiar presence

Corporations are not individual they have no rights

Existential is all inclusive to being

The movie; "Water World" produced by Kevin Costner doesn't make sense there was ever enough water to cause a global flood

Noah's ark is a fable; water evaporates and condensates recycling the same quantity which has existed since the planets chemistry separated post the big bang...

You know love knowing the obstacles another life has overcome

Liberty is a natural right its loss the seeds to revolutions and then there is simply what governments do to protect the people

The 1960's were in revolution against despotism and an intellectual revival from the 1950's cold war era which caused its paranoia in the fear of nuclear holocaust much as did the war on terrorism 2001

Every occupation has its own linguistic function

There is no true communism it is economic capitalism

Self-value plays many roles in most determinations

Kung Fu harnesses any physics to motion and intuitive virtue

Authoritarian eye for an eye tooth for a tooth has no virtue and is an infectious disease conditions despotism in transient causes

What you think and what you say has overlapping thought in construction or portrait depiction

911 set in motion an economic surge 2003 ending 2008 in recession

Government representatives should hold degrees in economics and sociology

Much speculation conditions rhetorical thought

Seat belt laws usurp probable cause increases municipal revenues in bias economics

United States is a strange name for a country in what vindication

Immigration; the native Americans are the only indigenous people in the United States

Nature's consensual age is self-reliance

Christianity has a posterious covenant with philosophy

All planets have the same chemistry certifies the possibility of other intelligent beings in the universe, proximity to suns conditions evolution for organic life

Notes:

Section One Hundred Thirty

Most television is mindless programming postulating a collective mind

Feeling betrayed has a sense in abandonment postulating who knew who

Government' economy & democracy have an ambiguous association

Economic legislation guarantees no individual occupational sustenance

Adolescent friends who are domineering are insecure and best ignored

Subconscious is a file drawer of experience documented in magic ink which only appears when prompted

Being real exposes common vulnerabilities distorting masculine virtues

Procurement in economic sustainability has a real fear in survival

Challenging intuit individual perspective is any depth in perception to culture

The movie Fisher King express's the perversions of transgression in capitalism

Invasions in Iraq & Afghanistan in a war on terrorism only made things worse in the deaths of countless innocent Muslims

What is happening post 911 in intellectual waking is consistent with other periods in despotism IE 1930's and 1960's only more transparent due to the internet

Hand Held Media Devices are an ardent correspondent

Existentialism' physical awareness has perplexing accouterments

Vices are yeoman social chameleons

If you live in a rural area or a small town there is no economic expansion providing positions for new people

Children sense a sinking feeling when scolded

When you are arrested you feel a sense of shame as a sinking feeling is a transfer in aggression is transgression… "Beverly Hills Cop" Eddie Murphy wanted you to know this

People think you are stupid not believing in anything

Nihilist mind knows no common virtue in value to contribution

Certain drugs (cocaine/meth) cause paranoiac polarities in mental derangement

Capitalism is perverting there is no restitution where there is no record expunge-ment the only restitution is truth knowing any causes as obedience to symbolic authority has no intuitive juncture

There is no constitutional authority knowing cause reasons why freedom to expression exists

There is no collective mind in knowledge to culture has any origins to the invention of Gods

A world view is an existential presence to culture is dualism in the world looking back at us in expectation

Notes:

Section One Hundred Thirty One

A nation of Immigrants "America" has no indigenous union polarizes race creed color prejudice

Laws in sexuality set precedence in obscure promiscuity

Kung Fu knows no vengeance

There are two polarities' the social being and the economic, economic conditions the social

Lecture limits reflection in constructive discovery

In mind in practice there is no end goal

There is no intelligent design there is no preceptual operativeness to culture, awareness is a chameleon cradle in adaptation to environment

Ego has a lingering posterity in reflection

Social economic structure has any chameleon paradigm

Television news anchors would be philosophy majors unifying perspectives

Petrified wood is molecular mineral sedimentation in crystalized platelets cast in a mummified suspended state forming molecular elements over millenniums

Intuit virtues physics to culture

When you do anything deceptive you sense a dark spell of suspicion

If you think the world view has changed it is your own perception there is no collective conscious state of mind only what media impersonates

Fear in life is a real factor wanting to be successful you know it when you want a new Porsche when you are sixteen

There are things you say and there are things you want to say to people posturing reservations

There is a cyber space convention in the menagerie of human transfer of conscious experience

Tradition has a draw to a dream state of mind

Reality is the relations you have with people

There are things you say and there are things you want to say to people posturing reservations

Reality is any impression of the arts

Time is a door to a world of illusion from home

Family has an auspicious polarity to perfection

Time wants any separation from work and play

Existential is a real pull in survival fitting in to a world of posterity

Notes:

Section One Hundred Thirty Two

The real world is a force in adaptation to existential accessory material procurement

Adam and Eve there is no world view knows anyone's worldly perception

Consensual age transition knows a certain deception to envious independence

Free loader moocher makes one feel like a piggy bank leper

Stamina has a degrading affect when you are not in good physical condition

NBN, Network Broadcast News postulates an authority

Credit cards are a banker's bounty

Nothing is as it ever was in perceptions to time

Perfectionism has no resolve

Clothing has a constant presence

Sleeping naked has a sense of freedom

Laws and taxations are politician virtues

Capitalisms ladders have humiliating rungs

Equality has no idiosyncratic convention

Trouble shooting anything requires physical relativities in construal

Perception in age is relative to exposure and familiarity

God **FEARING** people, does god fear people

Occupation conditions time on treadmills

In hard times people pile up in unreserved sanctuaries

Indigenous culture shares consistent features

Aggression is conditioned by a paternal sense to dejection

The Garden of Eden had no expectation when Adam &
Eve had no clothes

Alcohol exposes mysterious dispositions

Authoritarianism has a strange symposium

In transient correspondence no one is who we imagine
meeting in person

Notes:

Section One Hundred Thirty Three

There is no pre conceptual operativeness to conscious experience there is no intelligent design there is no cause there is no authority consciousness is conditioned by culture in adaptation

Political hindsight in conflicting global diffidence is a politician's platform

There is no matrix architect the matrix is any mind to the means interdependence

The Land of Fredonia, Duck Soup' the thirties thru the seventies express the best mainstream arts' inflation now distorts any vision

Happy Days the fifties were harsh in transient cause living in the cold war's despotism

History lessons are scripted lecture postulating comprehension

A sense of the world as a child is subjection to innocence challenging intuits has an earnest trust to authority

There is no collective motion picture to time there is no comprehension to what is not experienced reasons why history lessons should not be graded

Defying senses cigarettes are the most perverted first drug distorting purity in health

Intuit decides moral in membership to regret

It's rough when you have no money and no one to fall back on

Watching Star trek on METV is a time machine if you were there

The earth is flatland the universe is three dimensions navigational orientation makes compasses obsolete

Material domestication has a homing device orientation

A foragers home is anywhere in the woods

Sincerity has a charm in humility

Immigration has any economic prejudice

Intellect wants any balance to drama

Companionship in marriage conditions character

The corporate morning Glee Club View is a superficial stratum impostering a collective state of mind

---

Pressure causes stress conditions chemistry leads to chemical dependence

The tongue is the most sensitive organ disposed to chemical addictions (nicotine) conditions awareness to craving in withdrawal

Judgment of others is a speculative survey to commonality in appreciation often has reservations

Political media sessions assign no amicable resolves to mingling cultures distinguishing ideological divergence

A commitment to change any constant has a shift in focus

Notes:

Section One Hundred Thirty Four

Aging appearance has an awkward consolation

Having someone in your life stations time postulates one mind in view

Memory recall has any personal attraction

Any world view is a selective cinematic feature's media impression

A collective social view has a superficial characterization to distinctive knowledge

In High School there is a sense of union someone is looking out for you "short lived" after graduation where isolation displaces social gathering

Any world view is a kaleidoscopic station

Any social view is an environmental condition in repetitive familiarity

Sexual desire is a temporal fixation of a carnal compulsion conditioning stress wanting physical release combined to social "dejection" influences "criminal intent" sexual abuse is a form of temporary insanity

Intuition asks many unanswered questions in repressive quantities

Foreign policy is a neighbor's indiscretion unless a fruit tree hangs over a fence

Teasing people is a superfluous provocation

Perspective is an idiosyncratic station

Drug addictions condition a sense of corporal reclusion

Global media is a center eye reigning world view

Doubt is a non-contextual connoisseur

Power in suggestion impersonates postulated calculation to cause

Being upset has any stern composure wanting "restitution"

The force in forage conventionality to a material servitude estrange natural instincts in survival

Time is perceptive experience in culture

Manipulation has many points of view

Cursive slang has a discerning persuasion

Enthusiasm is an ardent covenant

Liberty is any art to mind

I've heard gnats are a delicacy

Reserve is the best reward

Notes:

ingramcontent.com/pod-product-compliance
ning Source LLC
bersburg PA
V031812170526
7CB00001B/33